Universitext

Editors

F.W. Gehring
P.R. Halmos

Universitext

Editors: F.W. Gehring, P.R. Halmos

Booss/Bleecker: Topology and Analysis
Charlap: Bieberbach Groups and Flat Manifolds
Chern: Complex Manifolds Without Potential Theory
Chorin/Marsden: A Mathematical Introduction to Fluid Mechanics
Cohn: A Classical Invitation to Algebraic Numbers and Class Fields
Curtis: Matrix Groups, 2nd. ed.
van Dalen: Logic and Structure
Devlin: Fundamentals of Contemporary Set Theory
Edwards: A Formal Background to Mathematics I a/b
Edwards: A Formal Background to Higher Mathematics II a/b
Endler: Valuation Theory
Frauenthal: Mathematical Modeling in Epidemiology
Gardiner: A First Course in Group Theory
Godbillon: Dynamical Systems on Surfaces
Greub: Multilinear Algebra
Hermes: Introduction to Mathematical Logic
Hurwitz/Kritikos: Lectures on Number Theory
Kelly/Matthews: The Non-Euclidean, The Hyperbolic Plane
Kostrikin: Introduction to Algebra
Luecking/Rubel: Complex Analysis: A Functional Analysis Approach
Lu: Singularity Theory and an Introduction to Catastrophe Theory
Marcus: Number Fields
McCarthy: Introduction to Arithmetical Functions
Meyer: Essential Mathematics for Applied Fields
Moise: Introductory Problem Course in Analysis and Topology
Øksendal: Stochastic Differential Equations
Porter/Woods: Extensions of Hausdorff Spaces
Rees: Notes on Geometry
Reisel: Elementary Theory of Metric Spaces
Rey: Introduction to Robust and Quasi-Robust Statistical Methods
Rickart: Natural Function Algebras
Schreiber: Differential Forms
Smoryński: Self-Reference and Modal Logic
Stanisić: The Mathematical Theory of Turbulence
Stroock: An Introduction to the Theory of Large Deviations
Sunder: An Introduction to von Neumann Algebras
Tolle: Optimization Methods

Leonard S. Charlap

Bieberbach Groups and Flat Manifolds

Springer-Verlag
New York Berlin Heidelberg London Paris Tokyo

Leonard S. Charlap
Institute for Defense Analysis
Communications Research Division
Princeton, NJ 08540

AMS Subject Classification: 30F10, 30F35

Library of Congress Cataloging in Publication Data
Charlap, Leonard S.
Bieberbach groups and flat manifolds.
(Universitext)
Bibliography: p.
Includes index.
1. Riemann surfaces. 2. Riemannian manifolds.
3. Functions, Automorphic. 4. Bieberbach groups.
I. Title.
QA333.C46 1986 516.3′62 86-15615

Printed and bound by R.R. Donnelley & Sons, Harrisonburg, Virginia.
Printed in the United States of America.

9 8 7 6 5 4 3 2 1

ISBN 0-387-96395-2 Springer-Verlag New York Berlin Heidelberg
ISBN 3-540-96395-2 Springer-Verlag Berlin Heidelberg New York

To Ann

Preface

Many mathematics books suffer from schizophrenia, and this is yet another. On the one hand it tries to be a reference for the basic results on flat riemannian manifolds. On the other hand it attempts to be a textbook which can be used for a second year graduate course. My aim was to keep the second personality dominant, but the reference persona kept breaking out especially at the end of sections in the form of remarks that contain more advanced material. To satisfy this reference persona, I'll begin by telling you a little about the subject matter of the book, and then I'll talk about the textbook aspect.

A *flat riemannian manifold* is a space in which you can talk about geometry (e.g. distance, angle, curvature, "straight lines," etc.) and, in addition, the geometry is locally the one we all know and love, namely euclidean geometry. This means that near any point of this space one can introduce coordinates so that with respect to these coordinates, the rules of euclidean geometry hold. These coordinates are not valid in the entire space, so you can't conclude the space is euclidean space itself. In this book we are mainly concerned with *compact* flat riemannian manifolds, and unless we say otherwise, we use the term "flat manifold" to mean "compact flat riemannian manifold."

It turns out that the most important invariant for flat manifolds is the fundamental group. In fact, these spaces are "classified" by their fundamental groups. Furthermore, these groups turn out to have many interesting algebraic properties. These groups are the *Bieberbach groups* of the title. They are torsionfree groups which have a maximal abelian subgroup of finite index which is free abelian. They are groups which are a little bit more complicated than free abelian groups. In recent years the groups have come to be studied somewhat more than the manifolds themselves, which is why they were awarded the first place in the title.

As you might have already guessed, the main results here are due to

Bieberbach who published his work in about 1910. His work was the solution to Hilbert's 18th problem. The results can be summed up in three theorems. I'll state them very roughly and not at all the way Bieberbach thought of them. The first says that the fundamental group of a flat manifold is a Bieberbach group. You sometimes hear the theorem stated that a flat manifold is covered by a torus, which comes to the same thing. The second says that two flat manifolds with the same fundamental groups are the same (e.g. homeomorphic). The third says that up to an appropriate equivalence there are only finitely many flat manifolds in each dimension.

Bieberbach's point of view was that of rigid motions, i.e. those transformations of euclidean space that preserve distance. Our first chapter is devoted to these rigid motions and to the proofs of Bieberbach's three theorems in this framework. The second chapter gives some background so we can state these theorems in the framework of flat manifolds. The third chapter is concerned with outlining a scheme for classifying Bieberbach groups and (consequently) flat manifolds. The fourth chapter carries out this classification in a special case, namely the case when the index of the free abelian subgroup of the Bieberbach group is prime. The fifth chapter tells about the automorphisms of Bieberbach groups and flat manifolds.

Why do I think that this mathematics is peculiarly suited for a second year graduate course? First of all, it is relatively elementary. Students with the standard first year graduate program under their belts should be able to handle the material contained herein. I have tried to present reasonably complete versions of any results that are not contained in such a first year program.

The main reason I think this is "good rich stuff" is because of its interdisciplinary character. If you glance at the table of contents you can easily see what I mean. There are sections on differential topology, algebraic number theory, riemannian geometry, cohomology of groups, and integral representations. Too often graduate students believe that mathematics divides up into neat segments labeled "complex variables," or "algebraic topology," or "group theory," for example. Of course, they get this idea because they take courses in which the subject matter is divided up in precisely this fashion. When they start to work on a thesis problem, they are sometimes quite surprised to discover that a problem in one area may lead them into a quite different area of mathematics. I thought it might be advisable to show them this phenomenon relatively early in their graduate studies, and this is the main reason I have written this book.

Another property this book shares with a number of other mathematics books is that it is "complex" in the sense that it is in two volumes, the first one (this one) is real while the second one is currently imaginary. I had hoped to have included material on the Hochschild–Serre spectral sequence, on the cobordism, cohomology, and K–theory of flat manifolds, and on the spectrum of flat manifolds. After a while, it began to appear that the book was growing too long to serve its main purpose, namely being a text, so I decided to leave this more advanced material for a projected second volume.

Now I suppose I should give the usual admonishment about doing the exercises. You don't really *have* to do all the exercises to appreciate the material. You should at least read them. Some are more important than others. Some merely point to nice extensions of various concepts while others are crucial to the proof of important theorems. Concerning the exercises that are embedded in proofs, I have tried to select those parts of proofs which would be boring or repetitious and turned them into exercises. This will, I hope, make them a little more interesting and give you, the reader, some idea of whether you are really following the argument.

I am glad to be able to thank a number of institutions for their support in the writing of this book. The State University of New York at Stony Brook gave me a sabbatical leave during the Fall term of 1982 during which time I began the manuscript. Harvard University provided facilities and a congenial atmosphere during the academic year 1982-83. The Communications Research Division of the Institute for Defense Analyses graciously allowed me to finish the book in 1984 and 1985.

I would like to thank my collaborators, Al Vasquez and Han Sah, not only for the usual "important comments and interesting conversations," but also for solving many of the problems in the subject that I found too difficult. In addition some of the material in Chapter V is joint work with Han Sah and is appearing here for the first time. I would also like to thank my teacher Jim Eells who did not merely "suggest" that I study flat manifolds. In a course in differential geometry at Columbia, he "assigned" to me the task of learning all there was to know about flat manifolds. (Fortunately less was known then than is known now.) A number of friends and colleagues have read portions of the manuscript in various of its forms and made comments, some of which have saved me from making egregious errors. Among them are William Goldman, Norman Herzberg, David Lieberman, and David Robbins. Lance Carnes of PCTEX gave invaluable assistance in the type setting. In addition, I would like to thank Ann Stehney for meticulous proofreading, knowledgeable comments, and comprehensive support. It is unlikely this book would exist at all without her very considerable contribution. Finally I will take full responsibility for all remaining obfuscations and errors (egregious or not).

Leonard S. Charlap

Princeton, Fall 1985

Table of Contents

Chapter I. Bierberbach's Three Theorems

Chapter II. Flat Riemannian Manifolds

Chapter III. Classification Theorems

ChapterIV. Holonomy Groups of Prime Order

Chapter V. Automorphisms

Some Notes on Notation

This book was typeset in my home using PCTEX on a Tandy 2000 computer. This gave me vastly more freedom to use different notation than one would ordinarily get with a typewriter. It is quite possible that I have yielded to temptation and abused this freedom. There are a number of mathematical abbreviations and special symbols that I use in my own writing, and I have taken the liberty to use them here. Some are quite standard ($\forall, \exists$, iff, etc.), and I won't comment on them here. The others are quite obvious ("s.t.", "w.r.t.", etc.), so I won't say anything about them either.

I have had a great deal of difficulty (the right word is really "tsoris") deciding on a method of numbering equations, theorems, exercises, etc., and the method I finally chose is far from satisfactory to me. Equation numbers are in parentheses and are started over at the beginning of each chapter. All the rest are numbered with a decimal number (e.g. 3.4) where the part to the left of the decimal point indicates the section, and the part to the right is numbered consecutively within that section, e.g. Theorem 3.4 refers to the fourth theorem in the third section of the current chapter.

I would have preferred a "Garden State Parkway" system of numbering in which an object is numbered by the page number on which it appears. I had to give this up because of the fact the more than one exercise would frequently appear on a page.

Chapter I

Bieberbach's Three Theorems

1. Rigid Motions

We start with an exercise (the first of many).

Exercise 1.1: Find all maps $f : \mathbb{R}^n \to \mathbb{R}^n$ that preserve distance.

Now we give the answer. Recall that O_n is the group of $n \times n$ real matrices m with $m^{-1} = m^t$. O_n is called the *n-dimensional orthogonal group*. GL_n will denote the group of $n \times n$ real matrices with $\det(m) \neq 0$. GL_n is the *real general linear group*.

Definition 1.1: A *rigid motion* is an ordered pair (m, s) with $m \in O_n$ and $s \in \mathbb{R}^n$. A rigid motion acts on $\mathbb{R}^n$ by

$$(m, s) \cdot x = mx + s \quad \text{for } x \in \mathbb{R}^n. \tag{1}$$

We multiply rigid motions by composing them, so

$$(m, s)(n, t) = (mn, mt + s). \tag{2}$$

Exercise 1.2: Show (2) that is the right formula, and that under this law of composition, the set of rigid motions forms a group. (**Hint:** You will find $(m, s)^{-1} = (m^{-1}, -m^{-1}s)$.)

We denote the group of rigid motions (of $\mathbb{R}^n$) by $\mathcal{M}_n$. We can write a rigid motion (m, s) as an $(n + 1) \times (n + 1)$ matrix,

$$\begin{pmatrix} m & s \\ 0 & 1 \end{pmatrix},$$

and then the composition (2) will just be the usual matrix multiplication.

Of course, the answer to Exercise 1.1 is the rigid motions. Although we are mostly interested in rigid motions, there is a generalization of them which will give rise to some interesting examples and conjectures.

Definition 1.2: An *affine motion* is an ordered pair (m, s) with $m \in GL_n$ and $s \in \mathbb{R}^n$.

Affine motions act on $\mathbb{R}^n$ via (1) and form a group, $\mathcal{A}_n$, under (2).

Clearly $\mathcal{M}_n$ is a subgroup of $\mathcal{A}_n$. We give the following definitions for $\mathcal{M}_n$, but you can see that analogous ones work for $\mathcal{A}_n$, too.

Definition 1.3: Define a homomorphism $r : \mathcal{M}_n \to O_n$ by

$$r(m, s) = m. \tag{3}$$

If $\alpha \in \mathcal{M}_n, r(\alpha)$ is called the *rotational part* of α. We can also define the *translational part* of α by

$$t(m, s) = s, \tag{4}$$

but the map $t : \mathcal{M}_n \to \mathbb{R}^n$ is not a homomorphism.

We say (m, s) is a *pure translation* if $m = I$, the identity matrix. If π is a subgroup of $\mathcal{M}_n, \pi \cap \mathbb{R}^n$ will denote the subgroup of π of pure translations.

Recall that such a group π is *torsionfree* if $\alpha \in \pi$ and $\alpha^k = (I, 0) \Rightarrow \alpha = (I, 0)$.

Exercise 1.3: Show that $\pi \cap \mathbb{R}^n$ is a torsionfree abelian normal subgroup of π, and $\pi/(\pi \cap \mathbb{R}^n)$ is isomorphic to $r(\pi)$.

If π is a subgroup of $\mathcal{M}_n$ and $x \in \mathbb{R}^n$, the set $\pi \cdot x = \{\alpha \cdot x : \alpha \in \pi\}$ is called the *orbit* of x. We will be interested in those subgroups whose orbits are discrete sets of points in $\mathbb{R}^n$.

Definition 1.4: We say a subgroup π of $\mathcal{M}_n$ (or $\mathcal{A}_n$) is *discontinuous* if all the orbits of π are discrete.

Exercise 1.4: $\mathcal{M}_n$ has the topology (but not the group structure) of $O_n \times \mathbb{R}^n$.

i) Show that a discrete subgroup of $\mathcal{M}_n$ is closed.

ii) Show that a subgroup $\pi \subset \mathcal{M}_n$ is discontinuous iff it is a discrete subset of $\mathcal{M}_n$. (**Hint:** Use the fact that O_n is compact)

iii) Is ii) true if $\pi \subset \mathcal{A}_n$? (**Hint:** Consider the subgroup $\{(1/2^i, x) : i \in \mathbb{Z} \text{ and } x \in \mathbb{R}\}$ of $\mathcal{A}_1$ and look at any orbit.)

While we want subgroups whose orbits are discrete, we also want each of these orbits to span $\mathbb{R}^n$. The reason for this will become somewhat apparent when we look at examples. (Of course, the main reason is that we need it to prove nice theorems.)

Definition 1.5: We say a subgroup π of $\mathcal{M}_n$ (or $\mathcal{A}_n$) is *uniform* if $\mathbb{R}^n/\pi$ is compact where the *orbit space*, $\mathbb{R}^n/\pi$, is the set of orbits with the quotient (or identification) topology. We say π is *reducible* if we can find $\alpha \in \mathcal{A}_n$, so that $t(\alpha\pi\alpha^{-1})$ does not span $\mathbb{R}^n$. In other words, π is reducible if after an affine change of coordinates, all the elements of π have translational parts in a proper subspace of $\mathbb{R}^n$. If π is not reducible, we say π is *irreducible*.

Exercise 1.5: Show that $\pi \subset \mathcal{M}_n$ is uniform iff it is irreducible. Is this true if $\pi \subset \mathcal{A}_n$?

Bieberbach's theorems are about discrete (or discontinuous), uniform (or irreducible) subgroups of $\mathcal{M}_n$, but for geometric applications of his theorems (to flat manifolds), we require the subgroups to have an additional property.

Definition 1.6: We say a subgroup π of $\mathcal{M}_n$ (or $\mathcal{A}_n$) acts *freely* on $\mathbb{R}^n$ if $\exists x \in \mathbb{R}^n$ with $\alpha \cdot x = x \Rightarrow \alpha = (I, 0)$, i.e. the only element of π that leaves any point in $\mathbb{R}^n$ fixed is the identity rigid motion.

Proposition 1.1: *Let π be a subgroup of $\mathcal{M}_n$ (or $\mathcal{A}_n$). Then if π acts freely, π is torsionfree. Furthermore, if π is discontinuous and torsionfree, π acts freely.*

Proof: Suppose $\alpha \in \pi, \alpha \neq (I, 0)$, and $\alpha^k = (I, 0)$. Let x be arbitrary in $\mathbb{R}^n$, and set

$$y = \sum_{i=0}^{k-1} \alpha^i \cdot x.$$

Then $\alpha \cdot y = y$, so π does not act freely.

Now suppose $\exists x \in \mathbb{R}^n$ and $\alpha \in \pi, \alpha \neq (I, 0)$, s.t. $\alpha \cdot x = x$. "Move the origin to x." More precisely, conjugate π by (I, x). Let $\pi' = (I, x)\pi(I, -x)$. Then if $\alpha = (m, s)$, we see that $(I, x)(m, s)(I, -x) = (m, s - (I - m)x)$. Then $\alpha \cdot x = x \Rightarrow s = (I - m)x = 0$, so $(m, 0) \in \pi'$. Clearly if π is discontinuous, π' is also. Therefore m cannot be a rotation through an irrational angle, so $\exists k$ s.t. $(m, 0)^k = (m^k, 0) = (I, 0)$. But if the conjugate of α has finite order, so does α, and π is not torsionfree. $\square$

Example 1.1: If m is a rotation by an irrational angle, then the subgroup of $\mathcal{M}_n$ generated by $(m, 0)$ will be torsionfree, but will not act freely since every element will leave the origin fixed.

The next exercise is very important and leads to the interesting open problem following it.

Exercise 1.6: Let π be a subgroup of $\mathcal{M}_n$ that acts freely and discontinuously on $\mathbb{R}^n$. Show that the orbit space $\mathbb{R}^n/\pi$ is an n-dimensional manifold, i.e. it is Hausdorff and locally euclidean (see Definition 2.1 in Chapter II if you don't know what this is). (**Hint:** You will need an "ϵ" argument for the locally euclidean part, and an "$\epsilon/2$" argument for the Hausdorff part.)

Open Problem 1.1: Is Exercise 1.6 true if π is merely a subgroup of $\mathcal{A}_n$?

Remark: i) This problem seems quite difficult. The difficulty arises because GL_n, in contrast to O_n, is not compact, so complicated things may happen to orbits quite "far away" (cf. Examples 2.7 and 2.8 below). There is an unpublished sketch of a proof for dimension 2 due to N. Kuiper, but we do not know of any results in higher dimensions. This "result" seems to have been tacitly assumed in a number of papers on "locally affine manifolds."

ii) The term "properly discontinuous" appears frequently in the context we have been discussing. We have avoided it because it seems to mean different thing to different authors. The main idea is that a group G acts on a space X *properly discontinuously* iff X/G is Hausdorff. A definition which is appropriate for our purposes is that G acts properly discontinuously on X iff $\forall x, y \in X$ with x, y in different orbits, $\exists$ neighborhoods N_x, N_y s.t. $G \cdot N_x \cap N_y = \emptyset$. The main point of the above open problem is to show that a subgroup of $\mathcal{A}_n$ that acts discontinuously (discrete orbits) acts properly discontinuously.

Definition 1.7: We say π is *isotropic* if $\pi \cap \mathbb{R}^n$ spans $\mathbb{R}^n$, and π is *crystallographic* if π is uniform and discrete. If π is crystallographic and torsionfree in $\mathcal{M}_n$, we say π is a *Bieberbach* subgroup of $\mathcal{M}_n$.

Note the apparent difference between isotropic and irreducible (or uniform). Isotropic requires that the pure translations span, while roughly speaking, irreducible just requires the translational parts of the elements of π span. If π is not discrete, this is a real difference. Also if π is in $\mathcal{A}_n$ but not $\mathcal{M}_n$, isotropic is stronger than irreducible even if π is discontinuous, but if π is a discrete uniform subgroup of $\mathcal{M}_n$, then it is isotropic.

This is Bieberbach's First Theorem. You may wonder at the term "crystallographic." While we don't want to get too involved (see [12] for a fuller discussion), we can introduce a notion here which will exhibit some crystalline shapes.

Definition 1.8: If $\pi \subset \mathcal{M}_n$ (or $\mathcal{A}_n$), a *fundamental domain* for π is a open subset $D \subset \mathbb{R}^n$ s.t. for each $x \in \mathbb{R}^n, \exists \alpha \in \pi$ s.t. $\alpha \cdot x \in \overline{D}$ and if $x, y \in D$ and $x \neq y, x$ and y are in different orbits, i.e. $\nexists \alpha \in \pi$ s.t. $\alpha \cdot x = y$.

If you look at some of the references mentioned in the next section, you will see that the fundamental domains for crystallographic groups are "crystalline."

One use of fundamental domains is to see what $\mathbb{R}^n/\pi$ looks like. Since each point in $\mathbb{R}^n$ is equivalent to one in $\overline{D}$, we can forget about $\mathbb{R}^n - D$. Also since distinct points in D are not equivalent, the only identifications that occur are on the boundary of D.

2. Examples

Example 2.1: Let $e_1, e_2, \ldots, e_n$ be the usual basis for $\mathbb{R}^n$. Let π_n be the subgroup of $\mathcal{M}_n$ generated by $(I, e_1), \ldots, (I, e_n)$. Then π_n is a Bieberbach subgroup of $\mathcal{M}_n$. $\mathbb{R}^n/\pi_n$ is the n-dimensional torus, T_n, which is homeomorphic to $S^1 \times S^1 \times \cdots S^1$.

A fundamental domain for π_n is just $\{(x_1, \ldots, x_n) \in \mathbb{R}^n : 0 < x_i < 1 \quad i = 1, 2, \ldots, n\}$. So for $n = 2$, we get

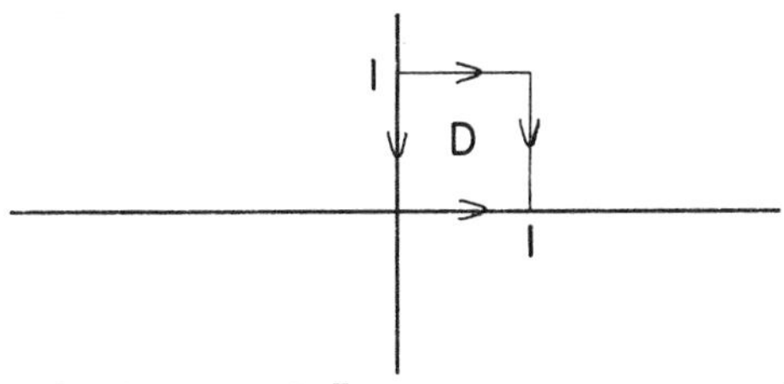

If we identify the top of $\bar{D}$ with the bottom of $\bar{D}$ according to the arrows, and the left side with the right side according to the arrows, we get a 2-torus.

Example 2.2: Let π be the subgroup (with two elements) generated by $(-I, 0)$ in $\mathcal{M}_1$. Clearly this is reducible. Now move the origin anywhere else, say to $x \neq 0$. Then $\pi' = (I, x)\pi(I, -x)$ will be generated by $(-I, 2x)$, so $T(\pi')$ spans $\mathbb{R}^1$, but π' is also reducible. Hence it is not enough to examine the set of translational parts to see if a group is reducible or not.

Example 2.3: Let $\pi \subset \mathcal{M}_2$ be generated by

$$\alpha = \begin{pmatrix} 1 & 0 & 1/2 \\ 0 & -1 & 1/2 \\ 0 & 0 & 1 \end{pmatrix} \text{ and } \beta = \begin{pmatrix} 1 & 0 & 0 \\ 0 & 1 & 1/2 \\ 0 & 0 & 1 \end{pmatrix}.$$

Exercise 2.1: Show π is Bieberbach and $\mathbb{R}^2/\pi$ is homeomorphic to the Klein Bottle.

It will follow from Theorem 7.1 of Chapter that up to an affine change of coordinates (i.e. conjugation in $\mathcal{A}_n$), this group and π_2 from Example 2.1 are the only Bieberbach subgroups of $\mathcal{M}_2$. You can, as a rather difficult exercise, try to prove this result here. It can be done with only elementary techniques. As a somewhat easier exercise, you can show that π is isotropic and that $r(\pi) \approx \mathbb{Z}_2$. In fact, π satisfies an exact sequence

$$0 \to \mathbb{Z} \oplus \mathbb{Z} \to \pi \to \mathbb{Z}_2 \to 1.$$

Also, show that π has the sole relation

$$\alpha\beta = \beta^{-1}\alpha, \tag{5}$$

and that every element of π can be written as $\beta^k\alpha^l$ for integers k and l, and finally, that

$$(\beta^k\alpha^l)(\beta^r\alpha^s) = \beta^{k+(-1)^l r}\alpha^{l+s}. \tag{6}$$

A fundamental domain for the Klein Bottle is pictured below.

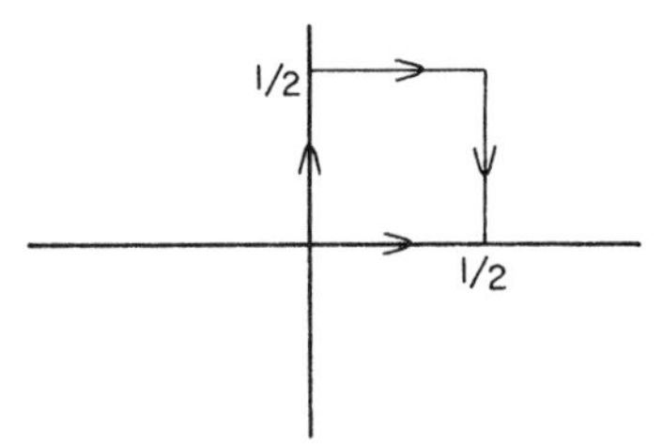

Example 2.4: Up to conjugation in $\mathcal{A}_3$, there are precisely ten Bieberbach subgroups of $\mathcal{M}_3$ (see [79] page 111). This was first proved by Hantsche and Wendt ([40]). The most interesting of these groups, sometimes called the Hantsche–Wendt group, is generated (in $\mathcal{M}_3$) by

$$\begin{pmatrix} -1 & 0 & 0 & 1/2 \\ 0 & -1 & 0 & 1/2 \\ 0 & 0 & 1 & 1/2 \\ 0 & 0 & 0 & 1 \end{pmatrix}$$

and

$$\begin{pmatrix} 1 & 0 & 0 & 1/2 \\ 0 & -1 & 0 & 0 \\ 0 & 0 & -1 & 0 \\ 0 & 0 & 0 & 1 \end{pmatrix}.$$

Call this group π.

Exercise 2.2: Show that π is a Bieberbach subgroup of $\mathcal{M}_3$, and π is isotropic, and $r(\pi)$ is isomorphic to $\mathbb{Z}_2 \oplus \mathbb{Z}_2$ (the Klein 4-group). Show that π satisfies an exact sequence

$$0 \to \mathbb{Z}^3 \to \pi \to \mathbb{Z}_2 \oplus \mathbb{Z}_2 \to 1.$$

A slightly more difficult result is that $\pi/[\pi, \pi] \approx \mathbb{Z}_4 \oplus \mathbb{Z}_4$.

It is this last fact that makes this π so interesting. It says, for example, that $\mathbb{R}^3/\pi$ has the same real homology as S^3, and with its canonical riemannian structure (see Chapter II) has no (global) Killing vector fields.

Example 2.5: This example is sometimes called the "infinite screw." Let $\pi \subset \mathcal{M}_3$ be generated by (m, e_3) where m is a rotation of irrational angle about the e_3-axis. It is trivial to see that π is discontinuous and acts freely on $\mathbb{R}^3$, but is not uniform. We have $\pi \cap \mathbb{R}^3 = \{(I, 0)\}$ and $r(\pi) \approx \mathbb{Z}(\approx \pi)$, so π is not isotropic, and $r(\pi)$ is not finite. Also since π is not uniform, π is not irreducible.

Example 2.6: This example is due to L. Auslander ([3]). It is a subgroup not of $\mathcal{M}_n$, but of $\mathcal{A}_n$, $\mathcal{A}_3$ in fact. It illustrates some more of the differences between $\mathcal{M}_n$ and $\mathcal{A}_n$. Let $\pi \subset \mathcal{A}_3$ be generated by

$$\alpha = \begin{pmatrix} 1 & 0 & 0 & 0 \\ 0 & 1 & 0 & 1 \\ 0 & 0 & 1 & 0 \\ 0 & 0 & 0 & 1 \end{pmatrix} \text{ and } \beta = \begin{pmatrix} 1 & 1 & 0 & 0 \\ 0 & 1 & 0 & 0 \\ 0 & 0 & 1 & 1 \\ 0 & 0 & 0 & 1 \end{pmatrix}.$$

Exercise 2.3: Show that π is a Bieberbach subgroup of $\mathcal{A}_3$ and $\mathbb{R}^3/\pi$ is a 3-manifold. (Be careful not to use Open Problem 1.) As a more difficult exercise, you can try to show that in spite of this, π is *not* isotropic. Notice that since π is discontinuous, if π were isotropic, $\pi \cap \mathbb{R}^3$ would be a three-dimensional lattice, and hence a free abelian subgroup of π of rank three, i.e. isomorphic to $\mathbb{Z}^3$. In fact, you can try to show that π has no free abelian subgroup of π of rank three. To do this you might let

$$\gamma = \begin{pmatrix} 1 & 0 & 0 & 1 \\ 0 & 1 & 0 & 0 \\ 0 & 0 & 1 & 0 \\ 0 & 0 & 0 & 1 \end{pmatrix},$$

and show that each $\xi \in \pi$ can be written

$$\xi = \gamma^r \alpha^s \beta^t = \begin{pmatrix} 1 & t & 0 & r \\ 0 & 1 & 0 & s \\ 0 & 0 & 1 & t \\ 0 & 0 & 0 & 1 \end{pmatrix},$$

and

$$(\gamma^r \alpha^s \beta^t)(\gamma^u \alpha^v \beta^w) = \gamma^{r+u+tv} \alpha^{s+v} \beta^{t+w}.$$

You can also show $r(\pi) \approx \mathbb{Z}$.

We have already remarked that the Bieberbach subgroups of $\mathcal{M}_2$ are (up to affine change of coordinates) just π_2 (corresponding to T_2) and the one corresponding to the Klein Bottle. If we look not at all Bieberbach subgroups of $\mathcal{A}_2$, but only at those subgroups π with $\mathbb{R}^2/\pi$ homeomorphic to T_2, even here the situation is extremely complex. These subgroups have been classified by N. Kuiper ([50]). We will give one example of them here.

Example 2.7: Let $\pi \subset A_2$ be generated by

$$\alpha = \begin{pmatrix} 1 & 1 & 0 \\ 0 & 1 & 1/2 \\ 0 & 0 & 1 \end{pmatrix} \text{ and, } \beta = \begin{pmatrix} 1 & q & \lambda \\ 0 & 1 & q \\ 0 & 0 & 1 \end{pmatrix}$$

where $q \in \mathbb{R}$ is irrational and $\lambda \in \mathbb{R}$ is s.t. $\lambda \neq \frac{1}{2}q(q-1)$ Then one can show that π is isomorphic to $\mathbb{Z} \oplus \mathbb{Z}$, and $\mathbb{R}^2/\pi$ is homeomorphic to T_2. Furthermore, an arbitrary element $\xi \in \pi$ can be written

$$\xi = \alpha^m \beta^n = \begin{pmatrix} 1 & m + nq & n\lambda + \frac{1}{2}q^2(n-1) + nmq + \frac{1}{2}m(m-1) \\ 0 & 1 & m + nq \\ 0 & 0 & 1 \end{pmatrix}.$$

Then it is not too difficult to see that π is Bieberbach and $\mathbb{R}^2/\pi$ is a 2-manifold homeomorphic to T_2, but although $r(\pi)$ is isomorphic to $\mathbb{Z} \oplus \mathbb{Z}$, it is not even closed in GL_2.

Example 2.8: This example is not even a subgroup of $\mathcal{A}_n$, but is included to show that **Open Problem 1** must depend crucially on properties of affine motions. The example was told to me by Paul Smith who credited it to Kerékjártó. It is an action of the integers on the plane. It suffices to say what the element "1" $\in \mathbb{Z}$ does to an abitrary point in the plane. To do this draw two parallel horizontal lines in the plane. Call them A and B. On A and above it, our transformation "1" sends everything one unit to the right. On B and below it, we send everything one unit to the left. Now, fill up the region between A and B by a family of curves that each approach A and B asymtotically, as pictured below.

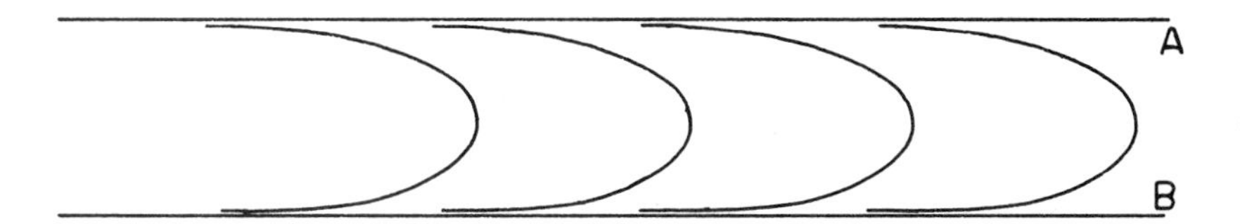

For example, if B is the x-axis, and A is the line $y = 1$, we could take for our family $x = \frac{1}{y(y-1)} + r$. Now in the region between A and B, each point will lie on precisely one curve, and "1" sends each point a distance of one unit along its curve. We illustrate the motion of some points below.

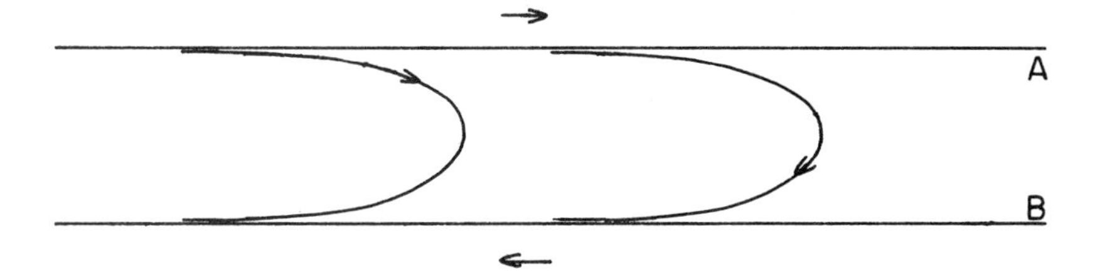

Exercise 2.4: Show that the transformation described above is continuous, and yields a discontinuous, free action of $\mathbb{Z}$ on $\mathbb{R}^2$. Show that the orbit space, $\mathbb{R}^2/\mathbb{Z}$, is locally euclidean, but not Hausdorff.

To compute $\mathbb{R}^2/\mathbb{Z}$, observe that the orbit of the region above and including A is a closed half cylinder, and so is the region below and including B. If you take a fundamental domain for the action in the region between A and B to be a narrow symmetric strip in the center, you will see that the orbit space of this is a whole (i.e. doubly infinite) cylinder. But no point in $A/\mathbb{Z}$ can be separated from any point in $B/\mathbb{Z}$.

Example 2.9: This example is a complement to the preceding one. In contrast, it *is* a subgroup of $\mathcal{A}_n$, but it does not act on all of $\mathbb{R}^2$. Let P be the punctured plane $\mathbb{R}^2 - \{0\}$, $\lambda > 1$ a real number, and let m be the matrix

$$\begin{pmatrix} \lambda & 0 \\ 0 & \lambda^{-1} \end{pmatrix}.$$

Let π be the (cyclic) group generated by m in $\mathcal{A}_n$.

Exercise 2.5: i) Show that π acts freely and discontinously on P but not on $\mathbb{R}^2$.

ii) Show that P/π is not Hausdorff. (**Hint:** Look at points on different coordinate axis.)

3. Bieberbach's First Theorem

The statement of Bieberbach's first theorem is easy. It says that a Bieberbach subgroup π of $\mathcal{M}_n$ is isotropic and $r(\pi)$ is finite. The difficult part of the proof is to show π contains one pure translation. The proof we give is modeled on the one in Wolf's book ([79]). After the proof, we will discuss some of the other proofs, in particular, Bieberbach's original one. Our proof proceeds by a series of lemmas.

Lemma 3.1: *There exists a neighborhood U of $I \in O_n$ s.t. $g \in O_n$ and $h \in U$ and $[g,[g,h]] = I \Rightarrow [g,h] = I$.*

Proof: If we assume that g and $[g,h]$ commute, then an easy computation shows that g and hgh^{-1} commute. Suppose that $\lambda_1, \ldots, \lambda_r$ are the distinct eigenvalues of g. Let V_i be the subspace of $\mathbb{C}^n$ generated by the eigenvectors of g corresponding to λ_i. We call $V_1, \ldots, V_r$ the *eigenspaces* of g. Notice that when restricted to V_i, g is a multiple (λ_i) of the identity.

Exercise 3.1: i) Let $m \in O_n$ and $W_1, \ldots, W_s$ be the eigenspaces of m. Show that $\mathbb{C}^n = W_1 \perp \cdots \perp W_s$ where "$\perp$" indicates orthogonal direct sum.

ii) Let $n \in O_n$. Show the eigenspaces of nmn^{-1} are $nW_1, \ldots, nW_s$.

iii) Suppose that $nW_i = W_i$ for $i = 1, 2, \ldots, s$. Show that n and m commute.

Since g and hgh^{-1} commute, we can pick a common set of eigenvectors for them, and they must have the same eigenspaces. On the other hand,

Exercise 3.1 shows that $hV_1, \ldots, hV_r$ are also the eigenspaces of hgh^{-1}. Since the eigenspaces are orthogonal, we must have $hV_i = V_j$ for some j. Since we can assume that h is near I, h cannot take a vector into one orthogonal to it. Hence $hV_i \cap V_j = \{0\}$ if $i \neq j$, and therefore $hV_i = V_i$ for $i = 1, 2, \ldots, r$. But $V_1, \ldots, V_r$ were also the eigenspaces for g, so by Exercise 3.1, g and h commute □

Lemma 3.2: *There exists a neighborhood V of $I \in O_n$ s.t. if $g, h \in V$, then the sequence*

$$g,\ [g,h],\ [g,[g,h]],\ [g,[g,[g,h]]],\ \ldots$$

converges to I in O_n.

Proof: Let $\|x\|$ be the length of x for $x \in \mathbb{R}^n$. For $m \in GL_n$, put

$$\|m\| = \sup\{mx : x \in \mathbb{R}^n \text{ and } \|x\| = 1\}.$$

So we have

$$\|mm'\| \leq \|m\| \cdot \|m'\|,$$

and

$$\|am + bm\pi\| \leq a\|m\| + b\|m\pi\| \quad \text{for } a, b \in \mathbb{R}.$$

Note that $m \in O_n \Rightarrow \|m\| = 1$. Now

$$\begin{aligned}\|I - [g,h]\,\| &= \|I - ghg^{-1}h^{-1}\| \leq \|hg - gh\| \\ &= \|(I-g)(I-h) - (I-h)(I-g)\| \leq 2\|I-g\| \cdot \|I-h\|.\end{aligned}$$

Putting $g_0 = g$, $g_1 = [g,h]$, $g_2 = [g,[g,h]]$, ..., we have

$$\|I - g_1\| \leq 2\|I-g\| \cdot \|I-h\|.$$

Hence

$$\|I - g_2\| \leq \|I-g\| \cdot \|I-g_1\| \leq 2^2\|I-g\|^2 \cdot \|I-h\|,$$

and, in general,

$$\|I - g_i\| \leq 2^i\|I-g\|^i \cdot \|I-h\| \qquad i = 1, 2, \ldots$$

Therefore, if we put $V = \{m \in O_n : \|I - m\| < \frac{1}{4}\}$, then

$$\|I - g_i\| \leq 2^i \left(\frac{1}{4}\right)^i \frac{1}{4} = \frac{1}{2^{i+2}},$$

and this $\to 0$ as $i \to \infty$. $\square$

Lemma 3.3: *Any neighborhood of $I \in O_n$ contains a neighborhood W of I s.t. $gWg^{-1} = W \quad \forall g \in O_n$.*

Proof: Put $W = \{m \in O_n : \|I - m\| < \epsilon\}$. Clearly for ϵ small enough, W will be contained in any given neighborhood of I. Now

$$\|I - gmg^{-1}\| \leq \|g - gm\| \leq \|g\| \cdot \|I - m\| = \|I - m\| \leq \epsilon \quad \forall g \in O_n.$$

Hence $m \in W \Rightarrow gmg^{-1} \in W$. $\square$

Exercise 3.2: For $\epsilon > 0$, let $B_\epsilon = \{m \in O_n : \|I - m\| < \epsilon\}$. Show that B_ϵ satisfies the conclusions of Lemmas 3.1, 3.2, and 3.3 if $\epsilon \leq \frac{1}{4}$.

The above three lemmas look somewhat technical (and they are), but remember we want to show that $\pi \cap \mathbb{R}^n$ is big (at least not empty), and that $r(\pi)$ is small (at least not infinite). These lemmas will be used to show that if $r(\pi)$ is big, π must have a large abelian subgroup which will tell us that $\pi \cap \mathbb{R}^n$ (which is, of course, abelian) is not small.

Lemma 3.4: *Let π be discrete in $\mathcal{M}_n$ and $A = (\overline{r(\pi)})_0$ be the identity component of the closure of $r(\pi)$ in O_n. Then A is abelian.*

Proof: Let $U \subset B_{\frac{1}{4}}$ be a neighborhood of I in O_n. Then by the above exercise, U satisfies the conclusions of the above three lemmas. Take $\gamma_1, \gamma_2 \in \pi$ s.t. $r(\gamma_1), r(\gamma_2) \in U$. Let $\gamma_{i+1} = [\gamma_1, \gamma_i]$ for $i \geq 2$. Write $\gamma_i = (g_i, t_i)$ with $g_i \in O_n$ and $t_i \in \mathbb{R}^n, i = 1, 2, \ldots$. Now

$$\gamma_{i+1} = [\gamma_1, \gamma_i] = ([g_1, g_i], g_1(I - g_i g_1^{-1} g_i^{-1}) \cdot t_i + (I - g_1 g_i g_1^{-1}) \cdot t_1). \quad (7)$$

So $g_{i+1} = [g_1, g_i]$, and we have

$$\|t_{i+1}\| \leq \|I - g_1^{-1}\| \cdot \|t_i\| + \|I - g_i\| \cdot \|t_1\|. \quad (8)$$

But $g_1 \in B_{\frac{1}{4}} \Rightarrow \|I - g_1^{-1}\| = \|g_1 - I\| = \|I - g_1\| < \frac{1}{4} \Rightarrow g_1^{-1} \in B_{\frac{1}{4}}$. So (8) yields

$$\|t_{i+1}\| \leq \frac{1}{4}\|t_i\| + \|I - g_i\| \cdot \|t_1\|. \quad (9)$$

By Lemma 3.3, $\lim_{i\to\infty} \|I - g_i\| \to 0$. Given $\epsilon > 0$, pick $N > 0$ s.t. $i > N \Rightarrow \|I - g_i\| \cdot \|t_1\| < \frac{1}{4}\epsilon$. Hence if $i > N$, (3) yields

$$\|t_{i+1}\| \leq \frac{1}{4}\|t_i\| + \frac{1}{4}\epsilon = \frac{1}{4}(\|t_i\| + \epsilon)$$

$$\Rightarrow \|t_{i+2}\| \leq \frac{1}{4}(\|t_{i+1}\| + \epsilon) \leq \frac{1}{4^2}\|t_i\| + \frac{1}{4^2}\epsilon + \frac{1}{4}\epsilon,$$

and in general,

$$\|t_{i+k}\| \leq \frac{1}{4^k}\|t_i\| + \left(\sum_{j=1}^{k} \frac{1}{4^j}\right)\epsilon. \tag{10}$$

Therefore we have $\lim_{i\to\infty} \|t_i\| \leq 0 + c\epsilon$ for some constant c and any $\epsilon > 0$. Hence $\lim_{i\to\infty} \|t_i\| = 0$ and $\lim_{i\to\infty} t_i = 0$.

Now we know $g_i \to I$ and $t_i \to 0$, so $\gamma_i \to (I, 0)$. But since π is discrete in $\mathcal{M}_n$, there must exist i_0 s.t. $\gamma_{i_0} = (I, 0)$. But by Lemma 3.1, this implies that for $i > 3$, then $g_i = I \Rightarrow g_{i-1} = I$, so $g_3 = [g_1, g_2] = I$.

Since g_1 and g_2 were arbitrary in $r(\pi) \cap U$, we see that $r(\pi) \cap U$ is abelian, and hence $\overline{r(\pi) \cap U}$ is abelian.

Exercise 3.3: Let G be a connected topological group. Show that any neighborhood of the identity generates the whole group. (**Hint:** Look at the subgroup generated by the neighborhood. Show it is open (easy) and closed (use the fact that cosets are open).)

We see that $(\overline{r(\pi)})_0 = A$ is generated by $\overline{r(\pi) \cap U}$ which is abelian, and the lemma follows. □

Lemma 3.5: *Let H be a closed subgroup of $\mathcal{M}_n$ with $\mathbb{R}^n \subset H_0$. Then H has only a finite number of components.*

Proof: Recall (or prove as an exercise) that the projection $p : H \to H/\mathbb{R}^n$ is continuous and open. So $H_0/\mathbb{R}^n$ is both open and closed in $H/\mathbb{R}^n$, and we must have $H_0/\mathbb{R}^n = (H/\mathbb{R}^n)_0$. Now by a standard isomorphism theorem,

$$H/H_0 \approx H/\mathbb{R}^n \Big/ H_0/\mathbb{R}^n \approx H/R^n \Big/ (H/\mathbb{R}^n)_0.$$

But $H/\mathbb{R}^n/(H/\mathbb{R}^n)_0$ is just the group of components of $H/\mathbb{R}^n$ in $M/\mathbb{R}^n$ which is O_n. So we are reduced to showing that a closed subgroup K of O_n can have at most a finite number of components.

Exercise 3.4: Show that O_n has a neighborhood of the identity which is homeomorphic to $\mathbb{R}^p$ where $p = n(n-1)/2$. (**Hint:** Look in Section II of Chapter I of [24]. You should know this material anyway.)

Since O_n is compact, if K has an infinite number of components, it will have an infinite number of components in the neighborhood of the identity mentioned in the exercise. Since K is supposed to be closed, it is easy to see that this can't happen. □

Except for Lemma 3.4, the above lemmas have been fairly general. The next lemma is very specific and contains the most important part of the proof. It describes an arbitrary discrete subgroup π of $\mathcal{M}_n$ in rather precise terms. The idea is to take an approximation π^* of π, and show that π^* is nice. To get π^*, we first "fatten up" π, and then take the intersection of π with the identity component of the "fat version" of π. So let

$$H = \overline{\pi \cdot \mathbb{R}^n}, \tag{11}$$

and

$$\pi^* = \pi \cap H_0 \tag{12}$$

where the closure in equation (11) is in $\mathcal{M}_n$. This H is the "fat" π. $\pi \cdot \mathbb{R}^n$ is just $\{(m, s) : m \in r(\pi) \text{ and } s \in \mathbb{R}^n\} = r(\pi) \cdot \mathbb{R}^n \subset \mathcal{M}_n$.

For example, if $r(\pi)$ is finite, then $\overline{\pi \cdot \mathbb{R}^n} = \pi \cdot \mathbb{R}^n$ and $H_0 = (\overline{\pi \cdot \mathbb{R}^n})_0$ is just $\mathbb{R}^n$ (or more precisely), $I \cdot \mathbb{R}^n$ so $\pi^* = \pi \cap \mathbb{R}^n$ is just the pure translations of π. On the other hand, if π is the infinite screw, then $\overline{\pi \cdot \mathbb{R}^3} = \{(m, s) : m \in SO_2 \text{ and } s \in \mathbb{R}^3\} = SO_2 \cdot \mathbb{R}^3$ where by SO_2 we mean the rotations about the x_3 (or z) axis. Since SO_2 is a circle, $H_0 = SO_2 \cdot \mathbb{R}^3$ and $\pi \cap H_0$ is just π itself.

If $r(\pi)$ is finite, the next lemma says π is isotropic. In the case of the infinite screw, the next lemma says π is isomorphic to the product of a finite abelian group (which is trival for the infinite screw) and a lattice.

Lemma 3.6: *Let π be a discrete subgroup of $\mathcal{M}_n$, and put $H = \overline{\pi \cdot \mathbb{R}^n}$ and $\pi^* = \pi \cap H_0$. Then*

i) π^* is normal in π and π/π^* is finite, and

ii) $\pi^* = A \times B$ where A is a finite abelian group and B is a finitely generated free abelian group of rank $\leq n$.

Proof: As remarked above,

$$\pi \cdot \mathbb{R}^n = \{(m, s) : m \in r(\pi) \text{ and } s \in \mathbb{R}^n\} = r^{-1}(r(\pi)). \tag{13}$$

This follows since $(m, x)(I, t) = (m, mt + x)$ and the fact that the equation $mt + x = s$ can always be solved for t.

So $\mathbb{R}^n \subset H_0$, and we can use Lemma 3.5 with $K = \mathbb{R}^n, H = H$, and $M = \mathcal{M}_n$ conclude that H_0 has finite index in H. Now H is the closure of

$\pi \cdot \mathbb{R}^n$, so each component of H contains an element of π, and $H = \pi \cdot H_0$. Therefore

$$\pi/\pi^* = \pi/\pi \cap H_0 \approx \pi \cdot H_0/H_0 \approx H/H_0, \tag{14}$$

which is finite, and i) follows.

Equation (13) implies $r(H_0) \subset (\overline{r(\pi)})_0$, therefore, by Lemma 3.4, $r(H_0)$ is abelian.

Exercise 3.5: Let G be a topological group and K a normal subgroup of G. Let H be any subgroup of G containing K. Let $p : G \to G/K$ be the canonical projection. Show $p(H)$ is closed and connected if H is closed and connected. (**Hint:** The connectedness is trivial. To see $p(H)$ is closed, compare G/H with $p(G)/p(H)$.)

Hence if we take $G = \mathcal{M}_n$, $K = \mathbb{R}^n$, and H in the exercise equal H_0, we can conclude that $r(H_0)$ is a closed connected subgroup of O_n since $\mathbb{R}^n \subset H_0$. Furthermore, since O_n is compact, $r(H_0)$ is compact. (At this point it is possible to see that $r(H_0)$ is a torus, but we won't need this fact.) Now let

$$W = \{x \in \mathbb{R}^n : g \cdot x = x \quad \forall g \in r(H_0)\} \tag{15}$$

and

$$\Delta = \pi^* \cap \mathbb{R}^n, \tag{16}$$

so Δ is the collection of pure translations in π^*.

Claim: $\Delta \subset W$.

First note that

$$[(m, s), (I, \delta)] = (I, m\delta - \delta). \tag{17}$$

Now take $\delta \in \Delta \subset \pi^* \subset \pi$, so $[\pi, \delta] \subset \pi$, and since π is discrete, $[\pi, \delta]$ is discrete. Since $[\mathbb{R}^n, \delta] = (I, 0)$, $[\pi \cdot \mathbb{R}^n, \delta]$ is discrete, and hence $[H, \delta]$ and $[H_0, \delta]$ are discrete. Since s does not appear on the right side of (17), $[H_0, \delta] = [r(H_0), \delta]$ where $r(H_0)$ is considered as a subgroup of $\mathcal{M}_n$ (i.e. g is identified with $(g, 0)$). Therefore $[r(H_0), \delta]$ is discrete, but since $r(H_0)$ is connected, so is $[r(H_0), \delta]$, and we must have $[r(H_0), \delta] = (I, 0)$. By (17),

$$[(g, 0), (I, \delta)] = (I, (g - I) \cdot \delta) = (I, 0), \qquad \forall g \in r(H_0).$$

Hence $g \cdot \delta = \delta, \forall g \in r(H_0)$, and the claim follows.

If $\gamma \in \pi^* \subset H_0$, by (15), W is left pointwise fixed by $r(\gamma)$. Pick $\gamma_0 \in \pi^*$ so that W is the maximal subspace of $\mathbb{R}^n$ left pointwise fixed by $r(\gamma_0)$. Say $\gamma_0 = (g_0, t_0)$. Write $\mathbb{R}^n = W \oplus W^\perp$. Since W is certainly g_0-invariant, so is $W^\perp$. Let $g^\perp = g_0 \mid W^\perp$ and write $t_0 = t + t^\perp$ with $t \in W$ and $t^\perp \in W^\perp$. Since W is the largest subspace of $\mathbb{R}^n$ left pointwise fixed by g_0, $g^\perp$ has no eigenvalues equal to 1, and $(I - g^\perp)$ is invertible. Let $x = (I - g^\perp)^{-1} \cdot t^\perp$, and as in Proposition 1, "move the origin to x," so $(g^\perp, t^\perp)$ goes to $(g^\perp, 0)$. So we can assume that $t^\perp = 0$ or $\gamma_0 = (g_0, t_0)$ with $t_0 \in W$.

Let $\gamma = (g, t)$ be arbitrary in π^*.

Claim: $t \in W$.

We have

$$\begin{aligned}[\gamma_0, \gamma] &= ([g_0, g], (g_0 - [g_0, g]) \cdot t + (I - g_0 g g_0^{-1}) \cdot t_0) \\ &= (I, (g_0 - I) \cdot t) \end{aligned} \tag{18}$$

since $g, g_0 \in r(H_0)$ which is abelian and since $(I - g) \cdot t_0 = 0$ because $t_0 \in W$. Hence $[\gamma_0, \gamma]$ is a pure translation in π^*, i.e. $[\gamma_0, \gamma] \in \Delta$. By the previous claim, $\Delta \subset W$, so $[\gamma_0, \gamma] \in W$.

Again write $\mathbb{R}^n = W \oplus W^\perp$ and $t = t' + t^\perp$. So $g_0 \cdot t - t = g_0 \cdot t^\perp - t^\perp \in W^\perp$. But (18) shows $(g_0 - I) \cdot t \in W$. Hence $g_0 \cdot t = t$ which implies $t \in W$, and the claim follows.

Now we know that $(g, t) \in \pi^* \Rightarrow g \in r(H_0)$ and $t \in W$, hence $\pi^* \subset r(H_0) \cdot W \subset \mathcal{M}_n$. But $r(H_0)$ consists of pure rotations, and W consists of pure translations, so $r(H_0) \cap W = (I, 0)$. Furthermore, (17) and the second claim show $[r(H_0), W] = (I, 0)$. Therefore, $r(H_0) \cdot W$ is a direct product, $r(H_0) \times W$, and r is the projection onto the first factor. Let

$$p : r(H_0) \times W \to W$$

be the projection onto the second factor.

Let $A = \ker(p|\pi^*) = \pi^* \cap r(H_0)$. Since $r(H_0)$ is compact, and π^* is discrete, A is finite. Since $r(H_0)$ is abelian, A is abelian. Since A is finite, $p(\pi^*)$ is discrete in W. Now a discrete subgroup of a finite dimensional vector space is a lattice, i.e. $p(\pi^*)$ is a free abelian group of rank $\leq n$.

We have $A = \ker(p|\pi^*)$ is finite and $\text{img}(p|\pi^*)$ is torsionfree Hence $\pi^* = A \times B$ where B is isomorphic to $p(\pi^*)$. $\square$

There is a tricky point in the last paragraph of this proof. B is isomorphic to $p(\pi^*)$, but $B \neq p(\pi^*)$. So even though $p(\pi^*) \subset W \subset \mathbb{R}^n, B$ may not consist of pure translations. For example, if π is the infinite screw, $\pi = \pi^* = B$, and it is true that π is a free abelian group of rank ≤ 3. W, in this case, is the z-axis, and p is just the map to the translational part.

Theorem 3.1 (First Bieberbach): *Let π be a crystallographic subgroup of $\mathcal{M}_n$. Then*

i) $r(\pi)$ is finite, and

ii) $\pi \cap \mathbb{R}^n$ is a lattice (finitely generated free abelian group) which spans $\mathbb{R}^n$, i.e. π is isotropic.

Proof: Naturally, we apply Lemma 3.6 to π. Since π is uniform and π/π^* is finite, π^* must be uniform. So there exists a compact fundamental domain $F \subset \mathbb{R}^n$ for π^*. This $\Rightarrow \exists m > 0$ s.t. $\forall x \in \mathbb{R}^n\ \exists \gamma \in \pi^*$ s.t.

$$\|\gamma \cdot x\| \leq m. \tag{19}$$

Let $p(\pi^*)$ be the subgroup of $\mathbb{R}^n$ as defined in Lemma 3.6, and let V be the subspace of $\mathbb{R}^n$ generated by $p(\pi^*)$.

Claim: $V = \mathbb{R}^n$.

Suppose not. Let $x \in V^\perp$ with $\|x\| = 2m$. Suppose $\gamma = (g, s) \in \pi^*$. Then $s \in V$. Now $\|g \cdot x\| = \|x\|$ since $g \in O_n$, and since $g \cdot x \in V^\perp, \|g \cdot x + s\| \geq \|x\|$ (the length of the sum of perpendicular vectors is always $\leq$ the length of either). In other words, $\|\gamma \cdot x\| \geq \|x\| = 2m$ for any $\gamma \in \pi^*$ which contradicts (19), and the claim follows.

Recall $[r(H_0), W] = (I, 0)$ and

$$[(g, 0), (I, t)] = (I, g \cdot t - t).$$

So $g \cdot t = t\ \forall g \in r(H_0)$ and $t \in W$. Therefore, $r(H_0)$ acts trivially on V which is all of $\mathbb{R}^n$, and $r(H_0)$ is trivial itself. Thus $\pi^* = B$ and since $r(H_0) = I$, this time B is contained in $\mathbb{R}^n$, and we see that $\pi^* = \pi \cap \mathbb{R}^n$ is the subgroup of pure translations of π. Now B (and π^*) spans $V = \mathbb{R}^n$, so $\pi \cap \mathbb{R}^n$ spans $\mathbb{R}^n$, and we already know that $\pi/\pi^* = \pi/\pi \cap \mathbb{R}^n$ is finite, and we are done. $\square$

Remarks: i) Recall that a sequence of groups and homomorphisms is *exact* if the image of any of the homomorphisms is equal to the kernel of the next one. If π is any crystallographic group, then π satisfies an exact sequence

$$0 \longrightarrow M \longrightarrow \pi \longrightarrow \Phi \longrightarrow 1 \tag{20}$$

where $M = \pi \cap \mathbb{R}^n$ is a lattice of rank n, and $\Phi = r(\pi)$ is a finite group. We call Φ, the *holonomy group* of π. We use the "0" at the left of (20) to indicate that we usually write M additively while the "1" at the right indicates that we usually write Φ mutiplicatively.

We will later see that if a group satisfies (20), it can be imbedded as a crystallographic subgroup of $\mathcal{M}_n$.

ii) A set of generators for $\pi \cap \mathbb{R}^n$ will be a basis for $\mathbb{R}^n$. If we write the matrix of any $g \in r(\pi)$ w.r.t. this basis, then this matrix will have integer entries, although, in general, it will no longer be orthogonal.

Part of the proof of Lemma 3.6 is important enough to be broken out as a lemma itself.

Lemma 3.7: *Let $(m, s) \in \mathcal{M}_n$. Then by possibly moving the origin, we can assume $m \cdot s = s$.*

4. Bieberbach's Second Theorem

Bierberbach's Second Theorem tells us that a crystallographic subgroup of $\mathcal{M}_n$ fits inside $\mathcal{M}_n$ in essentially only one way. This theorem is much easier to prove than the First Theorem and depends mostly on the algebraic structure of crystallographic groups, although this will not be so apparent from the proof we give here. The first part of the proof is an algebraic characterization of the pure translations of π.

Proposition 4.1: *Let π be a crystallographic subgroup of $\mathcal{M}_n$. Then $\pi \cap \mathbb{R}^n$ is the unique normal, maximal abelian subgroup of π.*

Proof: Let $\rho \subset \pi$ be a normal abelian subgroup. It suffices to show $\rho \subset \mathbb{R}^n$, i.e. $r(\rho) = I$. Let $(m, s) \in \rho$. By Lemma 3.7, we can assume $m \cdot s = s$. Let (I, v) be any element of $\pi \cap \mathbb{R}^n$. Then ρ normal implies

$$(I, v)(m, s)(I, -v) = (m, v - m \cdot v + s) \in \rho.$$

But ρ abelian $\Rightarrow [(m, s), (m, v - m \cdot v + s)] = (I, 0)$, which yields $(m - I) \cdot (v - m \cdot v + s) = (m - I) \cdot s$ and $(m - I) \cdot s = 0$, so we get

$$m \cdot v - m^2 \cdot v + m \cdot s - v + m \cdot v - s = 0,$$

or $(m - I)^2 \cdot v = 0$.

As before, let $X = \{x \in \mathbb{R}^n : m \cdot x = x\}$ and write $\mathbb{R}^n = X \oplus Y$ orthogonally. Write $v = v_x + v_y$ with $v_x \in X$ and $v_y \in Y$, then

$$(m - I)^2 \cdot v = (m - I)^2(v_x + v_y) = (m - I)^2 \cdot v_y = 0.$$

Now on Y, $m - I$ is non-singular, so $(m - I)^2 \cdot v_y = 0 \Rightarrow v_y = 0$. Hence $m \cdot v = v \quad \forall v \in \pi \cap \mathbb{R}^n$. Since $\pi \cap \mathbb{R}^n$ spans $\mathbb{R}^n$, $m \cdot x = x \quad \forall x \in \mathbb{R}^n$, so $m = I$. □

Theorem 4.1 (Second Bieberbach): *Let π and π' be crystallographic subgroups of $\mathcal{M}_n$. Let $f : \pi \to \pi'$ be an isomorphism. Then $\exists \alpha \in A_n$ s.t. $f(\beta) = \alpha\beta\alpha^{-1} \quad \forall \beta \in \pi$, i.e. any isomorphism between crystallographic groups can be realized by an affine change of coordinates.*

Proof: By Proposition 4.1, $\pi \cap \mathbb{R}^n$ and $\pi' \cap \mathbb{R}^n$ are the unique normal, maximal abelian subgroups of π and π' respectively. Hence $f(\pi \cap \mathbb{R}^n) = \pi' \cap \mathbb{R}^n$. Since $f \mid \pi \cap \mathbb{R}^n$ is an isomorphism between lattices of rank n in $\mathbb{R}^n$, $f \mid \pi \cap \mathbb{R}^n$ induces a linear map $g \in GL_n$.

Exercise 4.1: Show $r \circ f(m, s)$ is independent of s.

So we can define a map $f_1 : r(\pi) \to O_n$ by $f_1(m) = r \circ f(m, s)$ for any $(m, s) \in \pi$. Letting $f_2 : \pi \to \mathbb{R}^n$ be the other coordinate map, we have

$$f(m, s) = (f_1(m), f_2(m, s)) . \tag{21}$$

Now $(m, s)(I, u)(m, s)^{-1} = (I, m \cdot u)$. Taking f of both sides and using (21), we get $[f_1(m)g] \cdot u = (gm)u$, so by replacing u by $g^{-1}u$, we see that

$$f_1(m) = gmg^{-1}. \tag{22}$$

Define $F : \pi \to \mathcal{M}_n$ by

$$F(\alpha) = (g, 0)\alpha(g, 0)^{-1} \quad \forall \alpha \in \pi,$$

and $G : F(\pi) \to \pi'$ by $G = f \circ F^{-1}$, so G can be written, $G(m, s) = (m, G_2(m, s))$. So now it suffices to find $x \in \mathbb{R}^n$ s.t.

$$G(\beta) = (I, x)\beta(I, -x) \quad \forall \beta \in F(\pi),$$

since then we will have

$$f(\alpha) = (g, x)\alpha(g, x)^{-1} \quad \forall \alpha \in \pi.$$

Let

$$\rho = \{(m, v) \in \mathcal{M}_n : \exists (m, s) \in F(\pi) \text{ s.t. } v = s - G_2(m, s)\}.$$

Exercise 4.2: Show ρ is a subgroup and that $r(\rho) = r(F(\pi))$ so $r(\rho)$ is finite.

Now

$$f \circ F^{-1}(I, u) = f(I, g^{-1}u) = (I, u),$$

so $G \mid F(\pi) \cap \mathbb{R}^n = I$. Hence if $(I, u) \in F(\pi)$, $u - G_2(I, u) = 0$, so the only pure translation in ρ is $(I, 0)$. By Exercise 4.2, ρ is finite.

Exercise 4.3: Let ρ be a finite subgroup of $\mathcal{M}_n$. Show $\exists x \in \mathbb{R}^n$ s.t.

$$(I, x)(m, s)(I, -x) = (m, 0) \quad \forall (m, s) \in \rho,$$

i.e. by moving the origin to x we can make all elements of ρ pure rotations. (**Hint:** Let x be the sum of the translational parts of all of the elements of π.)

Now $(m, v) \in \rho \Rightarrow v = s - G_2(m, s)$ for some $(m, s) \in F(\pi)$, and $(I, x)(m, v)(I, -x) = (m, 0) \Rightarrow (I - m) \cdot x + v = 0$. These yield

$$(I - m) \cdot x = G_2(m, s) - s,$$

Which implies

$$G_2(m, s) = (I - m) \cdot x + s \quad \forall (m, s) \in F(\pi),$$

so

$$G(m, s) = (m, G_2(m, s)) = (m, (I - m) \cdot x + s) = (I, x)(m, s)(I, -x),$$

and we are done. □

5. Digression — Group Extensions

Bieberbach's Third Theorem says that up to an affine change of coordinates there are only finitely many crystallographic subgroups of $\mathcal{M}_n$ for each n. It follows directly from the first two theorems by using some standard (although deep) results from group theory. We would like to make these results explicit in a manner that will facilitate a systematic approach to the classification of Bieberbach groups.

The key fact to be exploited is the exact sequence below. Recall that

$$0 \longrightarrow M \longrightarrow \pi \longrightarrow \Phi \longrightarrow 1 \tag{23}$$

is exact, where π is any crystallographic subgroup of $\mathcal{M}_n$, $M = \pi \cap \mathbb{R}^n$ is a lattice of rank n and hence M is isomorphic to $\mathbb{Z}^n$, and Φ is a finite group. If π is, in addition, Bieberbach, then we know π is torsionfree. The idea behind the proof of the Third Theorem is to show there are only finitely many isomorphism classes of groups π satisfying (23), and then the Second Theorem will tell us that, up to an affine change of coordinates, there are only finitely many crystallographic subgroups of $\mathcal{M}_n$.

An exact sequence such as (23) is called a *group extension*, or sometimes we say π is an extension of Φ by M. We think of π as being built up from M and Φ. For example, consider some different ways we can get a group π which satisfies

$$0 \longrightarrow \mathbb{Z} \oplus \mathbb{Z} \longrightarrow \pi \longrightarrow \mathbb{Z}_2 \longrightarrow 1. \tag{24}$$

If you did Exercise 2.1, you know that one is the Bieberbach subgroup of $\mathcal{M}_2$ whose orbit space is the Klein Bottle. For the purpose of this discussion, let's call this group π_A. It is actually the most complicated of the groups which satisfy (24).

A somewhat simpler one is π_B which is generated by

$$\beta = \begin{pmatrix} 1 & 0 & 0 \\ 0 & 1 & 1 \\ 0 & 0 & 1 \end{pmatrix}, \quad \gamma = \begin{pmatrix} 1 & 0 & 1 \\ 0 & 1 & 1 \\ 0 & 0 & 1 \end{pmatrix}, \text{ and } \delta = \begin{pmatrix} 1 & 0 & 0 \\ 0 & -1 & 1 \\ 0 & 0 & 1 \end{pmatrix}.$$

In this case, the subgroup $\mathbb{Z} \oplus \mathbb{Z}$ is generated by β and γ, and the quotient group $\mathbb{Z}_2$ is generated by the projection of δ.

A third example, π_C, is generated by

$$\beta = \begin{pmatrix} 1 & 0 & 0 \\ 0 & 1 & 1 \\ 0 & 0 & 1 \end{pmatrix} \text{ and } \tilde{\gamma} = \begin{pmatrix} 1 & 0 & 1/2 \\ 0 & 1 & 1/2 \\ 0 & 0 & 1 \end{pmatrix},$$

and the subgroup $\mathbb{Z} \oplus \mathbb{Z}$ is generated by β and $(\tilde{\gamma})^2$. Since $\tilde{\gamma}$ is the only element of π_C not in the subgroup, its projection must generate the quotient group which is $\mathbb{Z}_2$ since $(\tilde{\gamma})^2$ is in the subgroup. Note that in π_A and π_B

the projection is just the map r which takes the rotational part, but in π_C the projection is different since $r(\pi_C)$ is trivial.

It is easily seen that π_C is isomorphic to $\mathbb{Z} \oplus \mathbb{Z}$ itself since it consists entirely of pure translations. We can take the free abelian subgroup of rank 2 to be $\mathbb{Z} \oplus 2\mathbb{Z}$, and the quotient will be, of course, $\mathbb{Z}_2$.

Finally, there is an obvious group, π_D, which fits into (24), namely the direct sum, $\mathbb{Z} \oplus \mathbb{Z} \oplus \mathbb{Z}_2$.

Exercise 5.1: Can π_D be imbedded in $\mathcal{M}_2$ (in any way)?

Let us observe that π_A, π_B, π_C, and π_D are all different, i.e. not isomorphic. π_A and π_B are not abelian, but π_C and π_D are. Also π_A and π_C are torsionfree, but π_B and π_D both have elements of order two. So no two of these groups are isomorphic, but each has a free abelian subgroup of rank 2 which has index 2.

We want to find all groups satisfying (23) or at least show there are only finitely many. The general theory of group extensions gives us a good start on this. We develop this theory below. You should keep examples π_A, π_B, π_C, and π_D in mind.

Definition 5.1: A *group extension* is an exact sequence

$$0 \longrightarrow K \xrightarrow{i} G \xrightarrow{p} Q \longrightarrow 1 \tag{25}$$

Of course, when we write such an exact sequence, we mean that K is isomorphic to a *normal* subgroup of G. For our purposes, it suffices to restrict to the case where K is abelian, and we include that as part of the definition. G and Q may be non-abelian. By an "abuse of terminology", we sometimes say that G is a group extension (of Q by K).

In examples π_A and π_B, the quotient group Q ($\approx \mathbb{Z}_2$) is represented by matrices, so if we think of $K \approx \mathbb{Z} \oplus \mathbb{Z}$ as the integral lattice in $\mathbb{R}^2$, we see that in this case, Q acts on K. Explicitly, in π_A,

$$\begin{pmatrix} 1 & 0 \\ 0 & -1 \end{pmatrix} \begin{pmatrix} 0 \\ 1 \end{pmatrix} = \begin{pmatrix} 0 \\ -1 \end{pmatrix} = - \begin{pmatrix} 0 \\ 1 \end{pmatrix},$$

and

$$\begin{pmatrix} 1 & 0 \\ 0 & -1 \end{pmatrix} \begin{pmatrix} 1 \\ 0 \end{pmatrix} = \begin{pmatrix} 1 \\ 0 \end{pmatrix}$$

where α^2 and β have been taken as the generators of $\pi \cap \mathbb{R}^n$. How can we get an action of Q on K in the general case?

If we look more closely at the rigid motion case, we can see that $(m,s)(I,u)(m,s)^{-1} = (I, m \cdot u)$, a fact we have used several times before. If $qv \in Q$, we can try setting

$$q \cdot k = gkg^{-1} \qquad \text{where } p(g) = q, \tag{26}$$

and $q \in Q, k \in K$, and $g \in G$. Since K is normal in $G, gkg^{-1} \in K$.

Exercise 5.2: Suppose $p(g') = q$ also. Show $g'k(g')^{-1} = gkg^{-1}$.

So (26) does define an action of Q on K. By the way, if we wanted to be pedantic, we would write (26) as

$$q \cdot k = i^{-1}(gi(k)g^{-1}).$$

We can see that in π_A and π_B, Q acts nontrivially on K, while in π_C and π_D, Q acts trivially on K. Hence the action of Q on K is not enough to determine the extension.

Definition 5.2: If K is an abelian group on which a group Q acts, we call K a *Q-module.*

Now suppose we are given Q and a Q-module, K. What can we do to find all extensions G, i.e. groups G which fit into (25) and yield the given Q-module structure on K under the action (26)? This question was solved in a straightforward manner a long time ago by the use of objects called *factor sets.* It turned out, however, that these notions gave rise to a beautiful theory of which they were merely a part. This theory is the cohomology of groups. Here we are just going to do the part which is useful for extensions, but we will do it in the terminology of the general theory. Later we will develop the general theory, and it will be very important in the study of Bieberbach groups. We start with a number of definitions.

Definition 5.3: A 2-*cochain* (on Q with coefficients in K) is any map $f : Q \times Q \to K$.

The map f has no algebraic properties at all. However, since 2-cochains take values in an abelian group, they form an abelian group too, i.e. $(f+g)(\sigma,\tau) = f(\sigma,\tau)+g(\sigma,\tau)$, etc. We denote this group by $C^2(Q;K)$.

We can define 3-cochains and 1-cochains (and in general j-cochains) in exactly the same manner, e.g., a 3-cochain is a map: $Q \times Q \times Q \to K$, etc. We thus get groups $C^3(Q;K)$ and $C^1(Q;K)$. We define homomorphisms

$$\delta^2 : C^2(Q;K) \to C^3(Q;K)$$

and

$$\delta^1 : C^1(Q;K) \to C^2(Q;K)$$

by the formulas

$$(\delta^2 f)(\sigma,\tau,\rho) = \sigma \cdot f(\tau,\rho) - f(\sigma\tau,\rho) + f(\sigma,\tau\rho) - f(\sigma,\tau) \qquad (27)$$

and

$$(\delta^1 g)(\sigma,\tau) = \sigma \cdot g(\tau) - g(\sigma\tau) + g(\sigma) \qquad (28)$$

for $f \in C^2(Q;K), g \in C^1(Q;K)$, and $\sigma,\tau,\rho \in Q$. δ^1 and δ^2 are the *coboundary* maps. Their definitions clearly involve not only the algebraic structure of Q and K, but also the action of Q on K. The formulas (27) and (28) may seem somewhat mysterious now, but they are imposed by the group axioms on G.

We call an element f in the kernel of δ^2 (i.e. satisfying $\delta^2 f = 0$) a 2-*cocycle*, and an element f in the image of δ^1, a 2-*coboundary*. We denote the group of 2-cocycles by $Z^2(Q;K)$, and the group of 2-coboundaries by $B^2(Q;K)$. Clearly Z^2 and B^2 are subgroups of C^2, but more is true.

Exercise 5.3: Show B^2 is a subgroup of Z^2. Equivalently, show that $\delta^2 \circ \delta^1 = 0$.

This computation is a prototype for many computations in homological algebra and algebraic topology. So we can define the *second cohomology group* (of Q with coefficients in K) by

$$H^2(Q;K) \stackrel{\text{def}}{=} Z^2(Q;K)/B^2(Q;K).$$

While there are many techniques for computing $H^2(Q;K)$, they only work in special cases. In general, we can say that the computation of $H^2(Q;K)$ is a distinctly non-trivial matter.

We are interested in $H^2(Q;K)$ because it is related to the extensions G of Q by K which yield the given action of Q on K. We would like to say that these extensions are in one-to-one correspondence with the elements of $H^2(Q;K)$, but unfortunately this is not the case, almost, but not quite. This slight difference will cause us no little grief when we try to classify Bieberbach groups, but it is not very important in the proof of the third theorem. We need one more definition in order to say what actually is true.

Definition 5.4: Let $0 \longrightarrow K \longrightarrow G \longrightarrow Q \longrightarrow 1$ and $0 \longrightarrow K \longrightarrow G' \longrightarrow Q \longrightarrow 1$ be two extensions. We say these extensions are *equivalent* if there exists a homomorphism $F : G \to G'$ s.t.

$$\begin{array}{ccccccccc} 0 & \longrightarrow & K & \longrightarrow & G & \longrightarrow & Q & \longrightarrow & 1 \\ & & \downarrow & & \downarrow F & & \downarrow & & \\ 0 & \longrightarrow & K & \longrightarrow & G' & \longrightarrow & Q & \longrightarrow & 1 \end{array}$$

is communative where the unlabeled vertical arrows are identity maps.

Exercise 5.4: i) Show F must be an isomorphism.

ii) Suppose G is isomorphic to $\mathbb{Z}_9$, the integers modulo 9. Let g be a generator. Let K be the subgroup of G generated by g^3, so K is isomorphic to $\mathbb{Z}_3$. Define two maps p and $p^* : G \to K$ by $p(g^i) = g^{3i}$ and $p^*(g^i) = g^{6i}$, e.g., $p(g^2) = g^6$, but $p^*(g^2) = g^3$. Then p and p^* define extensions

$$0 \longrightarrow K \longrightarrow G \xrightarrow{p} Q \longrightarrow 1 \tag{29}$$

and

$$0 \longrightarrow K \longrightarrow G \xrightarrow{p^*} Q \longrightarrow 1. \tag{30}$$

These are clearly extensions with the same quotient and kernel, and it is easy to see that the action of Q on K is the same. The groups in the middle are isomorphic, but you can (and should) show that the extensions (29) and (30) are not equivalent. So we see that equivalent extensions are isomorphic, but isomorphic extensions may not be equivalent.

Now we can state the desired theorem.

Theorem 5.1 : *If Q is any group and K is a Q-module, then the set of equivalence classes of extensions*

$$0 \longrightarrow K \longrightarrow G \longrightarrow Q \longrightarrow 1$$

is in one-to-one correspondence with the elements of the abelian group $H^2(Q;K)$.

Before giving the proof (which will be mostly a series of exercises anyway), we would like to look at our examples, π_A, π_B, π_C,and, π_D. Unfortunately, to use the theorem, we would need to know $H^2(Q;K)$ where $Q \approx \mathbb{Z}_2$ and $K \approx \mathbb{Z} \oplus \mathbb{Z}$ with two different actions. Let's look at the simplest non-trivial case we can imagine, a case even easier than the above. Take $Q = \mathbb{Z}_2$, and $K = \mathbb{Z}$, and let the action be trivial. Then even in this most elementary case, you will be hard pressed to show that $H^2(Q;K) \approx \mathbb{Z}_2$. Try it. It can be done in a straightforward manner if you are willing to work hard enough, but, of course, we will later have better non-straightforward methods to do it. Just so we can look a little at what *is* going on with our examples, let us state two propositions.

Proposition 5.1: *Let $Q = \mathbb{Z}_2$ be generated by q, and let K_1 be the Q-module $\mathbb{Z}$ with trivial action. Let K_2 be the Q-module $\mathbb{Z}$ with action given by $q \cdot 1 = -1$. Let K_3 be the Q-module $\mathbb{Z} \oplus \mathbb{Z}$ with $q \cdot (m,n) = (n,m)$. Then*

$$H^2(Q;K_1) \approx \mathbb{Z}_2, \tag{31}$$

$$H^2(Q;K_2) = 0, \text{ and} \tag{32}$$

$$H^2(Q;K_3) = 0. \tag{33}$$

This will follow from Proposition 4.2 of Chapter III.

If K is a Q-module, we say K is the *direct sum* of the Q-modules K_1 and K_2 (written $K = K_1 \oplus K_2$) if K is the direct sum of K_1 and K_2 as abelian groups, and $q \cdot (k_1, k_2) = (q \cdot k_1, q \cdot k_2)$ for $q \in Q$ and $k_1 \in K_1$, and $k_2 \in K_2$.

Proposition 5.2: *Suppose $K = K_1 \oplus K_2$. Then*

$$H^2(Q;K) \approx H^2(Q;K_1) \oplus H^2(Q;K_2).$$

Exercise 5.5: Prove this proposition. (**Hint:** Any map $f : S \to K_1 \oplus K_2$, where S is any set, is equivalent to two maps $f_1 : S \to K_1$ and $f_2 : S \to K_2$.)

In examples π_A, π_B, π_C,and π_D, Q is always $\mathbb{Z}_2$, and as an abelian group, K is $\mathbb{Z} \oplus \mathbb{Z}$, but there are two different actions of Q on K. In π_A

and π_B, the generator q of Q acts like the matrix

$$\begin{pmatrix} 1 & 0 \\ 0 & -1 \end{pmatrix},$$

while in π_C and π_D, Q acts trivially on K. Let us denote the first Q-module by K', and the second by K. Then Propositions 5.1 and 5.2 tell us that

$$H^2(Q; K) \approx \mathbb{Z}_2 \tag{34}$$

and

$$H^2(Q; K') \approx \mathbb{Z}_2 \oplus \mathbb{Z}_2. \tag{35}$$

Now we know π_A and π_B are not isomorphic, so clearly the extensions corresponding to them are not equivalent, and Theorem 5.1 and equation (34) tell us that we have discovered all the (inequivalent) extensions of Q by the Q-module K'. One of these extensions corresponds to the identity element of $H^2(Q; K)$, but which one, π_A or π_B? The answer to this question will follow easily from the proof of Theorem 5.1.

In the case of the trivial Q-module K, we know two inequivalent extensions, but according to (35), there are four inequivalent ones. Where are the other two? Are they isomorphic to either π_C or π_D? While it would be possible to answer these questions from the proof of Theorem 5.1, it would be messy. Later we shall prove a theorem (Theorem 2.2 of Chapter III) which will aid us in answering questions of this type. Now we give the proof of Theorem 5.1.

Proof: First let's assume we have an extension

$$0 \longrightarrow K \xrightarrow{i} G \xrightarrow{p} Q \longrightarrow 1, \tag{36}$$

and see how to get a cocycle $f : Q \times Q \to K$. Pick any map $s : Q \to G$ with the property that $p \circ s(q) = q$. The map s will not, in general, be a homomorphism and is certainly not unique. Such an s is sometimes called a *section* of (36) (or of G or of p). Now a general principle of homological algebra (which is what we are doing) is that when one comes across a map which is not a homomorphism, the function which measures how much it fails to be a homomorphism is going to be very interesting. Hence we want to look at elements of G of the form $s(q_1q_2) \cdot s(q_2)^{-1} \cdot s(q_1)^{-1}$. If s is a homomorphism, these will be the identity, but if it is not, they will not. So what we have is a map $f : Q \times Q \to G$ defined by

$$f(q_1, q_2) = s(q_1 q_2) \cdot s(q_2)^{-1} \cdot s(q_1)^{-1}.$$

But note that

$$p(s(q_1 q_2) \cdot s(q_2)^{-1} \cdot s(q_1)^{-1}) = q_1 q_2 q_2^{-1} q_1^{-1} = 1,$$

since p is certainly a homomorphism and $p \circ s$ is the identity map. Hence $f(q_1, q_2)$ is in K i.e. f maps into K. More precisely, we should say that $f(q_1, q_2)$ is in the image of i, but we think of K as a subgroup of G. Thus we have defined a 2-cochain $f : Q \times Q \to K$.

We want to show f is a 2-cocycle, i.e. $\delta^2 f = 0$. By (27), we get something like

$$\begin{aligned} \delta^2 f(q_1, q_2, q_3) = q_1 \cdot [s(q_2 q_3) s(q_3)^{-1} s(q_2)^{-1}] - s(q_1 q_2 q_3) s(q_3)^{-1} s(q_1 q_2)^{-1} \\ + s(q_1 q_2 q_3) s(q_2 q_3)^{-1} s(q_1)^{-1} - s(q_1 q_2) s(q_2)^{-1} s(q_1)^{-1}, \end{aligned} \tag{37}$$

and this doesn't look too promising. What is happening is that we are paying for our sloppiness in regarding K as a subgroup of G. Remember we are writing K additively and G multiplicatively, and (37) has these two operations all mixed up (this happens all too frequently in homological algebra). If we write out everything multiplicatively, and recall that $q \cdot k = s(q) \cdot k \cdot s(q)^{-1}$ by (26), we get

$$\begin{aligned} \delta^2 f(q_1, q_2, q_3) = &\left\{s(q_1) \cdot [s(q_2 q_3) s(q_3)^{-1} s(q_2)^{-1}] s(q_1)^{-1}\right\} \\ &\cdot \left\{s(q_1 q_2 q_3) s(q_3)^{-1} s(q_1 q_2)^{-1}\right\}^{-1} \\ &\cdot \left\{s(q_1 q_2 q_3) s(q_2 q_3)^{-1} s(q_1)^{-1}\right\} \\ &\cdot \left\{s(q_1 q_2) s(q_2)^{-1} s(q_1)^{-1}\right\}^{-1}. \end{aligned}$$

Now since K is abelian and the expressions in curly brackets are in K, we can permute them to our advantage. The expressions inside of the round brackets are in G and cannot be permuted. So we write the first expression, then the last one, then the second one, and finally the third.

$$\begin{aligned} \delta^2 f(q_1, q_2, q_3) &= s(q_1) s(q_2 q_3) s(q_3)^{-1} s(q_2)^{-1} s(q_1)^{-1} \\ s(q_1) s(q_2) s(q_1 q_2)^{-1} & s(q_1 q_2) s(q_3) s(q_1 q_2 q_3)^{-1} s(q_1 q_2 q_3) s(q_2 q_3)^{-1} s(q_1)^{-1} \\ &= 1 \end{aligned}$$

as desired. Notice we have used the associative law of G quite extensively.

Now we know that f defines a cohomology class in $H^2(Q;K)$, but how do we know that if we picked a different section, say $s' : Q \to G$, we wouldn't get a different cohomology class? We actually do get a different 2-cocycle, f', but it turns out that $f - f'$ is a coboundary. In fact, the definition of δ^1 is cooked up precisely to make this work.

Define f' analogously by

$$f'(q_1, q_2) = s'(q_1 q_2) \cdot s'(q_2)^{-1} \cdot s'(q_1)^{-1}.$$

We want a map (1-cochain) $g : Q \to K$ so that $\delta^1 g = f - f'$. The obvious function to try is $g(q) = s(q) \cdot s(q)^{-1}$.

Exercise 5.6: Show g is a 1-cochain (i.e. map): $Q \to K$ and that $\delta^1 g = f - f'$. Again you should write everything multiplicatively and in a shrewd order.

So now we have seen that an extension (36) determines a cohomology class $[f]$ in $H^2(Q;K)$. (We use $[\ ]$ to denote the coset of a 2-cocycle in $H^2 = Z^2/B^2$.) Now suppose we have a group Q, a Q-module K, and a cohomology class $\alpha \in H^2(Q;K)$. We want to get an extension G (or more precisely (36)).

We can save ourselves a lot of work if we pick a "nice" 2-cocycle $f \in \alpha$. "Nice" means $f(1,q) = f(q,1) = 0 \quad \forall q \in Q$ and is usually called *normalized.* First we must prove we can always get such a 2-cocycle in any cohomology class.

Lemma 5.1: *Let $f : Q \times Q \to K$ be any 2-cocycle. Then there is a normalized 2-cocycle f' in the same cohomology class as f.*

Proof: We must find $g : Q \to K$ so that $f' = f - \delta^1 g$ and f' is normalized. If you play around a little with formulas (27) and (28), you will see we want $g(q) = q \cdot f(1,1) \quad \forall q \in Q$. Then $\delta^1 g(q_1, q_2) = q_1 \cdot f(1,1)$ by (27), so

$$f'(q_1, q_2) = f(q_1, q_2) - q_1 \cdot f(1,1).$$

Now $f'(q,1) = f(q,1) - q \cdot f(1,1)$, but (27) implies

$$0 = \delta^2 f(q,1,1) = q \cdot f(1,1) - f(q,1) + f(q,1) - f(q,1),$$

so $f(q,1) = q \cdot f(1,1)$, and we get $f'(q,1) = 0 \quad \forall q \in Q$. In particular, $f'(1,1) = 0$.

Now $\delta^2 f' = \delta^2 f - \delta^2\delta^1 g = 0$, so (27) yields

$$0 = \delta^2 f'(1,1,q) = f'(1,q) - f'(1,q) + f'(1,q) - f'(1,1),$$

so $f'(1,q) = f'(1,1) \quad \forall q \in Q$. But we know that $f'(1,1) = 0$, so $f'(1,q) = 0 \quad \forall q \in Q$, and f' is normalized. □

Now we assume we have a normalized 2-cocycle $f \in H^2(Q;K)$. To find the extension G, we put $G = Q \times K$ as a set and define a multiplication on G by

$$(q,k)(q',k') = (qq', q \cdot k' + k + f(q,q')). \tag{38}$$

Exercise 5.7: Show that the multiplication as defined by (38) is associative. Observe that (27) is precisely what you need. In fact, (38) is associative if and only if f satisfies (27), and that is where (27) came from.

We would guess that $(1,0)$ is the identity of G, and since we are using a normalized 2-cocycle, this is indeed the case,

$$(1,0)(q,k) = (q, k + 0 + f(1,q)) = (q,k),$$
$$(q,k)(1,0) = (q, 0 + k + f(q,1)) = (q,k).$$

Inverses are also easy. By solving the obvious equation, we get

$$(q,k)^{-1} = (q^{-1}, -q^{-1} \cdot k - f(q^{-1},q)).$$

One side is easy to check,

$$\begin{aligned}(q^{-1}, -q^{-1} \cdot k - f(q^{-1},q)(q,k) =& (1, q^{-1} \cdot k - q^{-1} \cdot k \\ & - f(q^{-1},q) + f(q^{-1},q) \\ =& (1,0).\end{aligned}$$

To check the other side is a little trickier,

$$\begin{aligned}(q,k)(q^{-1}, -q^{-1} \cdot k - f(q^{-1},q) &= (1, -k - q \cdot f(q^{-1},q) + k + f(q,q^{-1})) \\ &= (1, f(q,q^{-1}) - q \cdot f(q^{-1},q)).\end{aligned}$$

Using the fact f is a normalized 2-cocycle, we get

$$0 = \delta^2 f(q,q^{-1},q) = q \cdot f(q^{-1},q) - f(1,q) + f(q,1) - f(q,q^{-1}),$$

and

$$f(q,q^{-1}) = q \cdot f(q^{-1},q). \tag{39}$$

Thus we see that G with the multiplication defined by (38) is a group. Now we must show that if we had picked a different normalized cocycle, $f' \in \alpha$, we would get an equivalent extension. Since f and f' are in the same cohomology class, there exists a 1-cochain $g : Q \to K$ s.t. $\delta^1 g = f - f'$. Let G' be the set $Q \times K$ with multiplication defined as in (38) with f' in place of f. Define $F : G \to G'$ by $F(q,k) = (q, k + g(q))$.

Exercise 5.8: Show F is an equivalence of extensions.

We now have correspondences in both directions, and must see that they are inverse to one another. Let's start with Q, a Q-module K, and $\alpha \in H^2(Q;K)$. We'll take a normalized 2-cocycle $f \in \alpha$ and form the extension G using (38). Then we do the other process and see if we get a cocycle in α. First we must choose a section $s : Q \to G$. Since $G = Q \times K$ as a set, the natural one to pick is $s(q) = (q,0)$. Then the 2-cocycle f' corresponding to s is defined by

$$\begin{aligned}
f'(q_1,q_2) &= s(q_1q_2)s(q_2)^{-1}s(q_1)^{-1} \\
&= (q_1q_2,0)(q_2,0)^{-1}(q_1,0)^{-1} \\
&= (q_1q_2,0)(q_2^{-1},-f(q_2^{-1},q_2))(q_1^{-1},-f(q_1^{-1},q_1)) \\
&= (q_1,-q_1q_2f(q_2^{-1},q_2)+f(q_1q_2,q_2^{-1}))(q_1^{-1},-f(q_1^{-1},q_1)) \\
&= (1,-q_1f(q_1^{-1},q_1)-q_1q_2f(q_2^{-1},q_2)+f(q_1q_2,q_2^{-1})+f(q_1^{-1},q_1)) \\
&= (1,-q_1f(q_2^{-1},q_2)+f(q_1q_2,q_2^{-1}))
\end{aligned}$$

by (39) used twice. Now

$$0 = \delta^2 f(q_1,q_2,q_1^{-1}) = q_1 \cdot f(q_2,q_2^{-1}) - f(q_1q_2,q_2^{-1}) + f(q_1,1) - f(q_1,q_2).$$

Hence $f'(q_1,q_2) = f(q_1,q_2)$ where by an "abuse of terminology," we have identified $(1,k)$ with k.

Finally, if we start with G and get f by

$$f'(q_1,q_2) = s(q_1q_2)s(q_2)^{-1}s(q_1)^{-1},$$

and then form G' using f via (38), we must show G and G' are equivalent extensions. Then we will be done. Recall that

$$0 \longrightarrow K \xrightarrow{i} G \xrightarrow{p} Q \longrightarrow 1 \tag{40}$$

is exact, so the obvious way to map G to G' is

$$F(g) = (p(g), g \cdot (s \circ p(g))^{-1}).$$

Now $g \cdot (s \circ p(g))^{-1}$ is in K (really $i(K)$) since

$$p(g \cdot (s \circ p(g))^{-1}) = p(g) \cdot (p \circ s)(p(g))^{-1} = 1.$$

Also if $g \in K$, then $F(g) = (1, g)$. It is also easy to see that F induces the identity on Q.

Exercise 5.9: Show F is an isomorphism (and hence an equivalence of extensions).

So the correspondence between extensions and cohomology classes is one-to-one, and the theorem follows. □

Notice that since $H^2(Q; K)$ is a group, there must be a corresponding way to add extensions. There is, and it is called *Baer addition* (or *multiplication*), but since we have no need for it, we won't go into it (see [57] page 69 if you are interested).

There is one particular extension which will be interesting to us. It is the one corresponding to the zero element of $H^2(Q; K)$. By looking at the above proof, we see that this extension G is the set $Q \times K$ with the multiplication

$$(q, k)(q', k') = (qq', q \cdot k' + k). \tag{41}$$

Definition 5.5: Let Q be a group and K a Q-module. The set $Q \times K$ with the multiplication (41) is called the *split extension* of Q by K or the *semi-direct product* of Q qnd K.

Exercise 5.10: If

$$0 \longrightarrow K \xrightarrow{i} G \xrightarrow{p} Q \longrightarrow 1 \tag{42}$$

is any extension, show it is split iff there exists a section $s : Q \to G$ which is a homomorphism. We say s *splits* (42).

Exercise 5.11: i) Show π_B and π_D are split, while π_A and π_C are not split.

ii) Show that $\mathcal{M}_n$ is the split extension of O_n by $\mathbb{R}^n$.

Now we see from (34) that there are only two extensions corresponding to the action

$$\begin{pmatrix} 1 & 0 \\ 0 & -1 \end{pmatrix}$$

of $\mathbb{Z}_2$ on $\mathbb{Z} \oplus \mathbb{Z}$, and we see that π_A corresponds to the non-zero element of $H^2(Q;K)$, while π_B corresponds to the zero element. In the case of the trivial action of $\mathbb{Z}_2$ on $\mathbb{Z} \oplus \mathbb{Z}$, π_D corresponds to the zero element. Notice that if Q acts trivially on K, the split extension is merely the direct product. We are still left with the question of identifying the three extensions which correspond to the three non-zero elements of $H^2(Q;K)$. You could figure this out directly from the proof of Theorem 5.1, but it would be messy.

Now recall we are trying to prove the third Bieberbach theorem which states (roughly) that there are only finitely many crystallographic subgroups of $\mathcal{M}_n$. We are first trying to show that there are only finitely many groups π which satisfy an exact sequence

$$0 \longrightarrow M \longrightarrow \pi \longrightarrow \Phi \longrightarrow 1 \tag{43}$$

where M is free abelian of rank n, and Φ is finite. If we can show $H^2(Q;K)$ is finite, we will have completed the first step, and we will have effected a significant reduction in our problem. For then, we will merely (!) be left with showing there are only finitely many groups Φ which have Φ-modules of rank n, and only finitely many such Φ-modules. We are being a bit sloppy here. We can clearly add to Φ any group, let it act trivially on M, but it will turn out that for crystallographic groups, no non-trivial element of Φ can act trivially on all of M, so that's not a problem. Right now we want to show that $H^2(Q;K)$ is finite.

Exercise 5.12: Let Q be a finite group and K a finitely generated Q-module. Show $H^2(Q;K)$ is a finitely generated abelian group.

Proposition 5.3: *Suppose the order of Φ is k. Then $k\alpha = 0 \quad \forall \alpha \in H^2(Q;K)$.*

Proof: Let f be any 2-cocycle in $H^2(Q;K)$ and define a 1-cochain g by

$$g(\sigma) = \sum_{\lambda \in \Phi} f(\sigma, \lambda).$$

Hence

$$\delta^1 g(\sigma, \tau) = \sigma \cdot \sum_{\lambda} f(\tau, \lambda) - \sum_{\lambda} f(\sigma\tau, \lambda) + \sum_{\lambda} f(\sigma, \lambda).$$

Since f is a cocycle, $\delta^2 f = 0$, so

$$f(\sigma\tau, \lambda) = \sigma \cdot f(\tau, \lambda) + f(\sigma, \tau\lambda) - f(\sigma, \tau),$$

and since we are summing over all of Φ,

$$\sum_{\lambda} f(\sigma, \tau\lambda) = \sum_{\lambda} f(\sigma, \lambda),$$

and we get

$$\begin{aligned}
&\delta^1 g(\sigma, \tau) \\
&= \sigma \cdot \sum_{\lambda} f(\tau, \lambda) - [\sigma \cdot \sum_{\lambda} f(\tau, \lambda) + \sum_{\lambda} f(\sigma, \lambda) - \sum_{\lambda} f(\sigma, \tau)] + \sum_{\lambda} f(\sigma, \lambda) \\
&= kf(\sigma, \tau).
\end{aligned}$$

Thus for any $f \in Z^2(\Phi; M)$, $kf \in B^2(\Phi; M)$. □

Corollary 5.1: *i) If Φ is finite, so is $H^2(\Phi; M)$.*

ii) If $(1/k) \cdot M \subset M$ (i.e. if for all $m \in M$, $\exists m' \in M$ s.t. $km' = m$), then $H^2(\Phi; M) = 0$, where k is the order of Φ.

The proofs of these statements are easy, and will be left as exercises.

6. Digression — Integral Repesentations of Finite Groups

The next step in our rather long-winded proof of Bieberbach's Third Theorem is to examine those finite groups Φ which have Φ-modules of rank n. Actually we are only concerned which Φ-modules M which are free as abelian groups. You probably know about modules over rings, and are perhaps wondering if there is a ring hiding somewhere here. The answer is yes, and Φ-modules are nothing more than modules over this ring.

Definition 6.1: Let Φ be a group. Then the *integral group ring of Φ*, denoted by $\mathbb{Z}[\Phi]$, is the set of all formal linear combinations

$$\sum_{\sigma \in \Phi} a_\sigma \sigma$$

where $a_\sigma \in \mathbb{Z}$ and only finitely many $a_\sigma \neq 0$. (This last condition is automatic if Φ is finite.) Addition and multiplication are defined as follows:

$$\sum a_\sigma \sigma + \sum b_\sigma \sigma = \sum (a_\sigma + b_\sigma) \sigma$$

and

$$\left(\sum a_\sigma \sigma\right) \left(\sum b_\sigma \sigma\right) = \sum c_\sigma \sigma$$

where

$$c_\sigma = \sum_{\tau \cdot \rho = \sigma} a_\tau b_\rho,$$

i.e. polynomial multiplication.

Exercise 6.1: Show $\mathbb{Z}[\Phi]$ is indeed a ring.

So a Φ-module is a module over the ring $\mathbb{Z}[\Phi]$, and the usual notions of homomorphism, isomorphism, direct sum, etc. of modules over rings apply here.

Convention: Unless explicitly stated otherwise, all Φ-modules considered here will be finitely generated and free abelian as groups.

Now suppose M is a Φ-module. We can choose a basis $m_1, m_2, \ldots, m_n$ for M as an abelian group since it is free. If $\sigma \in \Phi$, then σ corresponds to a matrix with respect to this basis, and this matrix will have integer entries. If we change the basis, the matrix for each σ will change by conjugation with some fixed matrix also with integer entries and determinent ± 1. Since $\sigma \in \Phi$ has an inverse, and since $\det(A^{-1}) = (\det A)^{-1}$, all the matrices corresponding to elements of Φ must have determinent ± 1, i.e. they must be in the group of $n \times n$ matrices with integer entries and determinent ± 1. This group is called the *unimodular group* and is denoted by J_n. Also, the matrix that corresponds to the change of basis must be in J_n. Therefore we can see that the notion of Φ-module is equivalent to the following notion:

Definition 6.2: Let Φ be a group. An *integral representation of Φ of rank n* is a homomorphism $R : \Phi \to J_n$. We say two such representations are *equivalent* if their images are conjugate in J_n. We say a representation is *effective* or *faithful* if R is injective. Effective representations correspond to *effective* modules, i.e. M is effective if

$$\{\sigma \in M : \sigma \cdot m = m \quad \forall m \in M\} = \{1\}.$$

Proposition 6.1: *Let π be a crystallographic subgroup of $\mathcal{M}_n$. Then the action of $\Phi = r(\pi)$ on $M = \pi \cap \mathbb{R}^n$ is effective.*

Exercise 6.2: Prove this proposition.

To get back to the proof of Bieberbach's third theorem, if we can show that J_n has only finitely many conjugacy classes of finite subgroups, then we will see that there are only finitely many isomorphism classes of effective Φ-module of rank n for all groups Φ. That ought to do it. To prove this we introduce an important auxilliary notion which is similar to one you may be familiar with.

Definition 6.3: Let M be a finitely generated free abelian group, i.e. a $\mathbb{Z}$-module. A *symmetric positive definite inner product* on M is a map

$$\rho : M \times M \to \mathbb{Z}$$

such that

i) ρ is bilinear, i.e. $\rho(m + m', n) = \rho(m, n) + \rho(m', n)$, etc.

ii) ρ is symmetric, i.e. $\rho(m, n) = \rho(n, m)$,

iii Given any homomorphism $\varphi : M \to \mathbb{Z}$, there exists a unique $n_0 \in M$ such that $\varphi(m) = \rho(m, n_0) \quad \forall m \in M$, and

iv) $\rho(m, m) > 0 \quad \forall m \neq 0$.

We will sometimes abuse the terminology and call ρ simply an inner product and M simply an inner product space. We say two inner product spaces (M, ρ) and (M', ρ') are *isomorphic* if there exists a group isomorphism $f : M \to M'$ s.t.

$$\rho'(f(m), f(n)) = \rho(m, n) \qquad (44).$$

To see what all this has to do with J_n, let $b_1, b_2, \ldots, b_n$ be a basis for M and $b'_1, b'_2, \ldots, b'_n$ be a basis for M'. Then ρ gives rise to a matrix B by $B_{i,j} = \rho(b_i, b_j)$.

Exercise 6.3: i) Show B is symmetric, invertible (and hence in J_n), and positive definite.

ii) If $U \in J_n$ is the matrix associated with the isomorphism $f : M \to M'$, show that (44) is equivalent to $B' = UBU^t$.

Definition 6.4: We say $U \in J_n$ is an *automorphism* of ρ if $B = UBU^t$, or equivalently, $\rho(U \cdot m, U \cdot n) = \rho(m, n)$ where by $U \cdot m$ we mean the usual automorphism on M induced by U with respect to $b_1, \ldots, b_n$. We let Φ_ρ be the subgroup of J_n consisting of all automorphisms of ρ.

Exercise 6.4: Let ρ and ρ' be inner products on M. Show ρ is isomorphic to ρ' iff Φ_ρ is conjugate to $\Phi_{\rho'}$ in J_n.

Now we see how inner product spaces are related to conjugacy classes of subgroups of J_n. The next proposition tells us where the finiteness comes in.

Proposition 6.2: *Let Φ be a subgroup of J_n. Then there exists an inner product ρ on $\mathbb{Z}^n$ s.t. Φ is a subgroup of Φ_ρ iff Φ is finite.*

Proof: Since the matrix B of ρ is symmetric, positive definite, and invertible, by a change of basis (or, if you prefer, by conjugation with an element of GL_n), we can assume that B is the identity matrix. Then $U \in \Phi_\rho$ means $UU^t = I$, or $U \in O_n$. But Φ_ρ is certainly discrete (it's in J_n), and O_n is certainly compact, so Φ_ρ must be finite.

Now suppose Φ is finite. Define an inner product on $\mathbb{Z}^n$ by

$$\rho(x, y) = \sum_{U \in \Phi} (Ux) \cdot (Uy)$$

where $x \cdot y$ is the usual inner product, $\sum x_i y_i$. Then $\Phi \subset \Phi_\rho$. $\square$

If we can show there are finitely many isomorphism classes of inner products on $\mathbb{Z}^n$, we will know what we need to know, and that happens to be a famous theorem of Eisenstein and Hermite and Jordan.

Theorem 6.1 : *There are only finitely many isomorphism classes of (symmetric positive definite) inner products on $\mathbb{Z}^n$.*

Proof: The proof is by induction on n. To begin, do the following:

Exercise 6.5: Show there is only one inner product on $\mathbb{Z}$.

The crux of the proof is the following lemma which we will later prove as a corollary to a famous theorem of Minkowski. To state this lemma, recall that a *lattice* L in $\mathbb{R}^n$ is a finitely generated (free abelian) subgroup of $\mathbb{R}^n$. If $b_1, \ldots, b_n$ is any basis for L, then the *volume* of L is the volume of a fundamental domain for L, i.e. $P = \{\sum r_i b_i : 0 \leq r_i \leq 1\}$ is a fundamental domain, and $\mathrm{vol}(P) = \int_P dx_1 dx_2 \cdots dx_n = \mathrm{vol}(L)$.

Exercise 6.6: Show

$$\mathrm{vol}(L) = |\det \mathrm{matrix}(b_1, \ldots, b_n)| = \sqrt{\det(b_i \cdot b_j)},$$

where $\mathrm{matrix}(b_1, \ldots, b_n)$ is the one whose rows are the vectors $b_1, \ldots, b_n$, and the second matrix $(b_i \cdot b_j)$ has entries $b_i \cdot b_j$ where the "$\cdot$" indicates the usual inner product on $\mathbb{R}^n$. Then show that vol(L) is independent of the choice of basis.

Exercise 6.7: Given an inner product ρ on $\mathbb{Z}^n$, show we can imbed $\mathbb{Z}^n$ as a lattice L in $\mathbb{R}^n$ s.t.

$$\rho(x, y) = x \cdot y, \text{ and}$$
$$\mathrm{vol}(L) = 1.$$

(**Hint:** If you know about tensor products, consider $\mathbb{Z}^n \otimes \mathbb{R}$, and if you don't, learn.)

So now we are reduced to showing that there finitely many isomorphism classes of lattices in $\mathbb{R}^n$ of volume one. Here is the promised lemma.

Lemma 6.1: *There is a number c_n with the property that any lattice $L \subset \mathbb{R}^n$ of volume one contains a point x_0 s.t.*

$$0 < x_0 \cdot x_0 \le c_n. \tag{45}$$

Now assuming the lemma and, by induction, that the theorem holds in dimensions less than n, let L be any lattice in $\mathbb{R}^n$ of volume one, and define a sublattice L_0 by

$$L_0 = \{y \in L : x_0 \cdot y \equiv 0 \bmod (x_0 \cdot x_0)\}$$

where x_0 is the point in Lemma 6.1.

Exercise 6.8: Show the index of L_0 in L is less than $x_0 \cdot x_0$.

Let $y \in L_0$. Then $(y \cdot x_0)/(x_0 \cdot x_0) \in \mathbb{Z}$ and $y - ((y \cdot x_0)/(x_0 \cdot x_0))\, x_0 \in L_0$. But

$$[y - ((y \cdot x_0)/(x_0 \cdot x_0))\, x_0] \cdot x_0 = 0,$$

hence L_0 is the orthogonal direct sum of $\langle x_0 \rangle$, the lattice generated by x_0, and $\langle x_0 \rangle^\perp$, the orthogonal complement of $\langle x_0 \rangle$ in L_0, i.e. $L_0 = \langle x_0 \rangle \perp \langle x_0 \rangle^\perp$. By induction, there are only finitely many possibilities for inner products for $\langle x_0 \rangle$ and $\langle x_0 \rangle^\perp$, so there are only finitely many possibilities for inner products of L_0, up to isomorphism.

By the last exercise and (45), the index of L_0 in L is bounded by c_n, so there are only finitely many possibilities for inner products for L. □

Corollary 6.1: *i) There are only finitely many conjugacy classes of finite subgroups of J_n.*

ii) There are only finitely many Φ-modules of rank n where Φ can be any finite group.

The corollary follows from the theorem and previous discussion. We are left with the proof of the lemma. Recall that a subset K of $\mathbb{R}^n$ is *convex* if $x, y \in K$ implies $\lambda x + (1-\lambda)y \in K \quad \forall \lambda \in [0,1]$. Also K is *symmetric about* 0 if $x \in K$ implies $-x \in K$.

Theorem 6.2 (Minkowski): *Let K be a convex subset of $\mathbb{R}^n$ which is symmetric about 0. Let L be any lattice in $\mathbb{R}^n$. If*

$$\mathrm{vol}(K) > 2^n \mathrm{vol}(L), \tag{46}$$

then K contains a non-zero point of L.

Proof: We have seen in Section 2 that $\mathbb{R}^n/L$ is a torus. Consider the projection $p : \mathbb{R}^n \to \mathbb{R}^n/L$. Let $K/2 = \{(x/2) : x \in K\}$. Clearly $\mathrm{vol}(K/2) = (1/2^n)\mathrm{vol}(K)$, so (46) yields

$$\mathrm{vol}(K/2) > \mathrm{vol}(L).$$

Hence the restriction of p to $K/2$ cannot be injective. There must be two distinct points $(x/2), (y/2) \in K/2$, with $p(x/2) = p(y/2)$, so $0 \neq (x/2) - (y/2) \in L$. But $(x/2) - (y/2)$ is the midpoint of the line between the two points x and $-y$ of K. ($-y \in K$ since K is symmetric.) Hence $(x/2) - (y/2)$ is in K and in L. □

Corollary 6.2: *Lemma 6.1 is true.*

Proof: Let $B(r) = \{x \in \mathbb{R}^n : |x| \leq r\}$ be the closed ball of radius r. Let

$$\frac{\mathrm{vol}(B(r))}{r^n} = \omega_n.$$

$B(r)$ clearly is convex and symmetric, so if $r^n > (2^n/\omega_n)$, then $B(r)$ contains a point of L (recall $\mathrm{vol}(L) = 1$ in Lemma 6.1).

Exercise 6.9: Show that if $r^n = (2^n/\omega_n)$, then $B(r)$ contains a point of L. (**Hint:** Look at $B(r+\epsilon)$ and let $\epsilon \to 0$.)

Therefore, every lattice of volume 1 contains a point x with $0 < x \cdot x \leq (2^n/\omega_n)^{2/n}$, and we can take

$$c_n = \frac{4}{(\omega_n)^{2/n}}.$$

□

Exercise 6.10: i) Show $\omega_1 = 2, \omega_2 = \pi, \omega_3 = \frac{4}{3}\pi, \omega_4 = \frac{\pi^2}{2}$, and

$$\omega_n = \int_0^{2\pi} \int_0^1 \omega_{n-2}(\sqrt{1-r^2})^{n-2} r \cdot dr\, d\theta = \frac{2\pi}{n}\omega_{n-2}.$$

Hence $\omega_n = \pi^{n/2}/(n/2)!$ if n is even. (In general, $\omega_n = \pi^{n/2}/\Gamma(n/2)$.)

ii) Show $c_n < 1 + \frac{1}{4^n}$. (**Hint:** This is not so easy. Compute c_n for $n < 12$. For $n \geq 12$, let

$$A_n = (1 + \frac{1}{4^n})^{n/2} \omega_n / 2^n,$$

and show by induction that $A_n > A_{n-2} > 1$.)

iii) Show

$$\lim_{n \to \infty} c_n \Big/ \left(\frac{2n}{\pi e}\right) = 1.$$

This treatment of inner product spaces has been taken from the exposition in [46].

7. Bieberbach's Third Theorem and Some Remarks

By now it should be apparent that Bieberbach's Third Theorem is quite easy given the results of the previous two sections which are all very well-known.

Theorem 7.1 (Bieberbach's Third): *Up to an affine change of coordinates, there are only finitely many crystallographic subgroups of $\mathcal{M}_n$.*

Proof: By Bieberbach's Second Theorem, it suffices to show there are only finitely many isomorphism classes of crystallographic subgroups of $\mathcal{M}_n$. By Bieberbach's First Theorem, every such crystallographic group π satisfies an exact sequence

$$0 \longrightarrow M \longrightarrow \pi \longrightarrow \Phi \longrightarrow 1 \tag{47}$$

where M is free abelian of rank n and Φ is finite and acts effectively on M. But Theorem 5.1 and Corollaries 5.1 and 6.1 say that there are on finitely many equivalence classes of extensions of this type. Since there are more equivalence classes of extensions than isomorphism classes of groups π, we are done. □

Remarks: i) The Third Theorem immediately leads to the question, "How many crystallographic or Bieberbach groups are there in dimension n?" The following table gives what is known:

dimension	1	2	3	4
number of crystallographic groups	1	17	219	4783
number of Bieberbach groups	1	2	10	74 .

We will have more to say about this problem later, but for now, we only want to point out that the results on crystallographic groups in dimension four required a high-speed computer (cf [12]), and unless some drastic new ideas are forthcoming, our main hope of carrying the results to higher dimensions rest on faster computers. A more fruitful approach is to fix the holonomy group Φ and ask for all Bieberbach groups π with $r(\pi) = \Phi$. We will take this path in later sections.

ii) The Second Theorem is really just a corollary of the First Theorem, while, as we remarked, the third follows immediately from the first two and standard material. So the meat all lies in the First Theorem. Except for Bieberbach's original proof ([9]) and a recent one by Peter Buser ([13] and [14]), all the proofs we have seen are variations on the proof of Frobenius ([34]) given shortly after Bieberbach's. Bieberbach's proof appears to make essential use of the following non-trivial result from number theory:

Theorem 7.2 : *Let $\theta_1, \ldots, \theta_l$ be real numbers. Then there exist integers $x_1, \ldots, x_l$ and n s.t.*

$$|\theta_i - \frac{x_i}{n}| < n^{-(1+1/l)} \qquad \textit{for } i = 1, 2, \ldots, l.$$

Basically this theorem tells us about the approximation of irrationals by rationals.

Peter Buser's new proof resulted from a study of the techniques that Gromov used in his work on almost flat manifolds ([38]). In fact, Gromov has said that his work on almost flat manifolds resulted from an attempt to understand what's really going on in the proof of Bieberbach's First Theorem.

iii) There has been a great deal of effort attempting to generalize Bieberbach's First Theorem to the affine group. In [59], Milnor conjectured that any subgroup of $\mathcal{A}_n$ that acts properly discontinuously must have a solvable subgroup of finite index. Recently Margulis gave an example of a sugroup π of $\mathcal{A}_3$ that acts properly discontinuously on $\mathbb{R}^3$ s.t. π is isomorphic to the free group on two generators. Margulis' example has non-compact quotient, so the conjecture is still open in the case we are interested in. See [33] for more information.

Chapter II

Flat Riemannian Manifolds

1. Introduction

Bieberbach groups can be studied purely for their algebraic properties which are extremely interesting. If we do so, however, we will be missing out on some wonderful results in riemannian geometry which, after a little initial work, come free with the algebraic results on Bieberbach groups. This procedure in which problems in one field, riemannian geometry, are converted to problems in another field, algebra, is very much in the spirit of modern mathematics.

In this chapter we will give the background necessary for the understanding of these geometric results. Since there are many good references for this material, we will give somewhat less detail than in the other chapters.

2. A Tiny Bit of Differential Topology

Definition 2.1: A *differential n-manifold* is a separable Hausdorff topological space, X, (if you prefer, think "separable metric space") together with a maximal collection $\{U_\alpha\}$ of open subsets and homeomorphisms $g_\alpha : U_\alpha \to V_\alpha$ (V_α open in $\mathbb{R}^n$) s.t. $X = \bigcup_\alpha U_\alpha$ and the composition

$$(g_\alpha \mid U_\alpha \cap U_\beta) \circ g_\beta^{-1} : V_\beta \to V_\alpha$$

is infinitely differentiable (i.e. smooth) $\forall \alpha, \beta$. This is sometimes called a *smooth* manifold. A function $f : U \to \mathbb{R}$, for U open in X, is said to be *smooth* if $(f \mid U \cap U_\alpha) \circ g_\alpha^{-1} : V_\alpha \to \mathbb{R}$ is smooth $\forall \alpha$. $C^\infty(U)$ denotes the vector space of smooth functions on U.

If $x \in U_\alpha$, then $g_\alpha(x) \in \mathbb{R}^n$, we can write $g_\alpha(x) = (g_\alpha^1(x), \ldots, g_\alpha^n(x)) = (x_1(x), x_2(x), \ldots, x_n(x))$, and these numbers are called the *coordinates of x with respect to U_α or g_α*. The pair (U_α, g_α) is called a *coordinate system at x*.

Definition 2.2: Let $x \in X$. A *tangent vector* at x is a map V_x which assigns a real number to each smooth function $f : U \to \mathbb{R}$, where U is a neighborhood of x. The mapping V_x must satisfy the following:

$$i) V_x(af + bg) = aV_x(f) + bV_x(g) \quad \text{for } a, b \in \mathbb{R}, \text{ and}$$
$$ii) V_x(fg) = f(x)V_x(g) + V_x(f)g(x).$$

Note that if f is defined on U and g is defined on U', then $f + g$ and fg are defined on any open subset of $U \cap U'$.

The first equation above says that tangent vectors are linear, while the second says they obey a product rule analogous to the Leibnitz rule for derivatives of functions. In fact, the number $V_x(f)$ is sometimes called the derivative of the function f in the direction V_x. The set of tangent vectors at x form, in the obvious fashion, a vector space called the *tangent space of* X at x and denoted by X_x or sometimes $T_x(X)$.

Exercise 2.1: i) Let (U_α, g_α) be a coordinate system at x and f a smooth function defined near x. Define

$$\left[\frac{\partial}{\partial x_i}(x)\right](f) = \left[\frac{\partial}{\partial x_i}(f \circ g_\alpha)\right](g_\alpha(x)).$$

Show that the map $\frac{\partial}{\partial x_i}(x)$ is a tangent vector at x.

ii) Show that the tangent vectors $\frac{\partial}{\partial x_1}(x), \ldots, \frac{\partial}{\partial x_n}(x)$ form a basis for X_x, so X_x has dimension n. (**Hint:** If $f : \mathbb{R}^n \to \mathbb{R}$ is smooth and $f(0) = 0$, you can find functions $g_i : \mathbb{R}^n \to \mathbb{R}$ s.t. $f(x) = \sum_{i=1}^n x_i g_i(x)$. (**Hint:** Write $f_x(t) = f(tx)$, so $f(x) = \int_0^1 f'_x(t)dt$.))

So, if you like, you can think of a tangent vector as being a linear combination of partial derivatives, and if you think of $\frac{\partial}{\partial x_i}$ as being an arrow of unit length pointing along the x-axis, you get the usual picture of a vector.

We can give still another interpretation of tangent vectors.

Definition 2.3: A *smooth curve* on X is a map $c : [0,1] \to X$ s.t. $g_\alpha \circ c$ is smooth for all α. (You have to worry about what it means for a function to be differentiable on a closed interval like $[0,1]$, but one interpretation that works is to assume that $g_\alpha \circ c$ can be extended so as to be differentiable on some open interval containing $[0,1]$.) A *broken* or *piece smooth* curve on X is a continuous map $c : [0,1] \to X$ with the property that $g_\alpha \circ c$ is piecewise smooth for all α, i.e. there is a finite decomposition of $[0,1]$ into subintervals, and $g_\alpha \circ c$ is smooth when restricted to each subinterval. Most of the time, it matters not whether we use smooth or broken curves. If we do not say which kind we are using, it should be safe to assume smooth.

If $c(t) = x$, we can define a tangent vector $\dot{c}_t$ at x by

$$\dot{c}_t(f) = \left[\frac{d}{dt}(f \circ c)\right](t).$$

$\dot{c}_t$ is called the *tangent vector along c at t or at x.*

An alternative definition of tangent vector defines two curves c and s to be *equivalent at x* if $c(t_1) = s(t_2) = x$ and $\dot{c}_{t_1} = \dot{s}_{t_2}$, i.e.

$$\left[\frac{d}{dt}(f \circ c)\right](t_1) = \left[\frac{d}{dt}(f \circ s)\right](t_2)$$

for all smooth f defined near x. Then a tangent vector at x is an equivalence class of curves.

Exercise 2.2: Show this is an equivalent definition.

The advantage of this alternative definition is that it works if you have an infinite dimensional manifold (which we never will).

Definition 2.4: Let U be open in X. A *(smooth) vector field V* on U is a map which assigns to each $x \in U$ a tangent vector $V_x \in X_x$ s.t. if $f \in C^\infty(U)$, the map $x \mapsto V_x(f)$ is also smooth.

If c is a curve on X, a *(smooth) vector field v along c* is a map which assigns to each $t \in [0,1]$ a vector $v_t \in X_{c(t)}$ s.t. the map $t \mapsto v_t(f)$ is a smooth map near $t \in [0,1]$ whenever f is smooth near $c(t) \in X$.

Exercise 2.3: Show that $t \mapsto \dot{c}_t$ is a vector field along c.

3. Connections and Curvature

Definition 3.1: Let $x \in X$. A *connection* ∇ *at* x is a map which assigns to a pair (U_x, V) a vector $\nabla_{U_x} V \in X_x$, where $U_x \in X_x$ and V is a vector field defined near x. ∇ must satisfy

i) the map $(U_x, V) \mapsto \nabla_{U_x} V$ is bilinear, and

ii) if f is smooth near x, then

$$\nabla_{U_x}(f \cdot V) = U_x(f) \cdot V_x + f(x) \cdot \nabla_{U_x} V. \tag{1}$$

A *connection on* X is a map which assigns to each $x \in X$ a connection at x s.t. if U and V are (smooth) vector fields, the map $x \mapsto \nabla_{U_x} V$ is a (smooth) vector field.

Remarks: i) Roughly speaking, $\nabla_{U_x} V$ measures how much away from parallel the vector field V is, along a curve in the direction of U_x at the point x. We'll make this more precise (or somewhat more precise anyway) below.

ii) You may have noticed that the conditions in the definition of connection are very similar to those in the definition of tangent vector. Mappings such as these which satisfy a linearity condition and a product rule are called *derivations*. Tangent vectors are derivations on the vector space of smooth functions considered as a vector space over the real numbers. A connection at a point x assigns to each tangent vector at x a derivation on the module of vector fields defined near x considered as a module over the ring of smooth functions defined near x.

Exercise 3.1: i) Show that $(U, V) \mapsto \nabla_U V$ is a bilinear map from pairs of vector fields to vector fields.

ii) Show that $\nabla_{fU} V = f \cdot \nabla_U V$.

iii) Show that $\nabla_U(f \cdot V) = U(f) \cdot V + f \cdot \nabla_U V$.

iv) If we have a coordinate system (U_α, g_α), we can write

$$\nabla_{\frac{\partial}{\partial x_i}} \frac{\partial}{\partial x_j} = \sum_k \Gamma^k_{i,j} \frac{\partial}{\partial x_k}$$

where the $\Gamma^k_{i,j}$ are functions defined in U_α. Show that the connection is completely determined in U_α by these functions. The $\Gamma^k_{i,j}$ are called the *Christoffel symbols* of ∇.

Definition 3.2: Suppose X and Y are differential manifolds. A map $F : X \to Y$ is said to be *smooth* if $f \in C^\infty(Y) \Rightarrow f \circ F \in C^\infty(X)$. If F is a homeomorphism, and F and F^{-1} are smooth, we say F is a *diffeomorphism.*

If $V_x \in X_x$, then we can define $dF_x(V_x) \in Y_{F(x)}$ by

$$[dF_x(V_x)](f) = V_x(f \circ F) \quad \text{for } f \in C^\infty(Y).$$

dF_x is called the *differential* of F at x.

Exercise 3.2: Show $dF_x : X_x \to Y_{F(x)}$ is linear.

Definition 3.3: Suppose that ∇ is a connection on Y and that $F : X \to Y$ is locally a diffeomorphism. We get a connection $F^*(\nabla) = \nabla^*$ on X by setting

$$\nabla^*_U(V) = \nabla_{dF(U)}(dF(V)),$$

where U and V are vector fields on X, and $dF(U)$ is the vector field on $F(X) \subset Y$ that sends f to $[dF_x(U_x)](f)$ where f is smooth near $F(x)$. ∇^* is called the *induced connection.* If X has a connection $\tilde{\nabla}$, $\nabla^* = \tilde{\nabla}$, and F is a diffeomorphism, we say F is an *affine equivalence.*

Now we are really interested in knowing when two tangent vectors, say V_x and U_y, at different points $x, y \in X$ are parallel. It turns out that, in general, we can't make any sense out of this problem because there isn't any natural way to get from x to y. In $\mathbb{R}^n$, there is the unique straight line from x to y, so we can say what parallel means, but for an arbitrary manifold X, we can't. The way around this difficulty is not to ask when U_x and V_y are parallel, but to ask when U_x and V_y are parallel along a particular curve c from x to y, i.e. $c(a) = x$ and $c(b) = y$ for some $a, b \in [0,1]$. What we want is a transformation (probably linear) $\|_a^b : X_{c(a)} \to X_{c(b)}$, and then if $\|_a^b (V_x) = U_y$, we would say V_x and U_x are parallel along c. We would call $\|_a^b$ *parallel translation along* c from a to b or from $c(a)$ to $c(b)$.

Suppose we knew how to do this, i.e. we had a linear transformation $\|_a^b : X_{c(a)} \to X_{c(b)}$ for any curve c and any $a, b \in [0,1]$. Then if we had a vector field V along c, we could define the *derivative of* V *along* c at some $t_0 \in [0,1]$ by

$$\frac{DV}{dt}\Big|_{t_0} = \lim_{h \to 0} \frac{1}{h} \left[\|_a^b (V_{t_0+h}) - V_{t_0} \right]. \tag{2}$$

$\frac{DV}{dt}$ is another vector field along c. We would say V is parallel along c if $\frac{DV}{dt}$ were 0 along c. Since we have ∇ and want $\|_a^b$, the way to do this is to reverse this procedure. An examination of (2) will show that $\frac{D}{dt}$ should satisfy

$$\frac{D}{dt}(V+W) = \frac{DV}{dt} + \frac{DW}{dt}, \tag{3}$$

$$\frac{D}{dt}(f \cdot V) = \dot{c}(f)V + f\frac{DV}{dt}, \tag{4}$$

and, if there exists a vector field $\tilde{V}$ on X with $V = \tilde{V}\mid_c$, then

$$\frac{DV}{dt} = \nabla_{\dot{c}}\tilde{V}. \tag{5}$$

Lemma 3.1: *There is a unique mapping $\frac{D}{dt}$ of vector fields along c to vector fields along c which satisfies (3),(4), and (5).*

Exercise 3.3: Prove this lemma. (**Hint:** Write everything in coordinates or see [61] page46.)

Thus given a connection ∇ and a curve c, the derivative along c is uniquely defined.

Lemma 3.2: *Given a curve c on X and a vector $V_{c(0)} \in X_{c(0)}$, there is a unique vector field V along c s.t. $V_0 = V_{c(0)}$ and which is parallel along c, i.e. $\frac{DV}{dt} = 0$ along c.*

Exercise 3.4: Prove this lemma. (**Hint:** Write everything in coordinates and use the basic existence and uniqueness theorem for differential equations, or see [61] page47.)

Why should this be called parallel? Suppose you have a manifold imbedded in $\mathbb{R}^n$. (Think of a surface in $\mathbb{R}^3$ if you like.) What would we like "parallel along a curve" to mean? We can't just move a tangent vector to the manifold along the curve keeping it parallel to itself in $\mathbb{R}^n$ because as we do this, it would pop out of the tangent space to the manifold. Let's imagine, however, that we have a vector $V_{c(0)} \in X_{c(0)}$, and we move it a little tiny bit to $c(\Delta t)$ by keeping it parallel in $\mathbb{R}^n$. Now it's no longer a tangent vector, but we can project it perpendicularly back into the tangent space to get a tangent vector. So to define the vector field V along c which is parallel in X with $V_0 = V_{c(0)}$, we can imagine a little person with a hammer riding on the vector, and as the vector tries to pop out of the tangent space, he or she continuously pounds it back in. (This image is due to Dirk Struik.)

Believe it or not, mathematical sense can be made of the previous paragraph, and it can be used to define a connection on X. What connection is it? Consider the inclusion mapping $i : X \to \mathbb{R}^n$. $\mathbb{R}^n$ has a natural connection defined by its parallelism; call it ∇. Then it is possible to show that the connection defined above (by the person with the hammer) is the induced connection $i^*(\nabla)$.

One of the many things you can do with a connection is to define an interesting invariant.

Definition 3.4: Let $x \in X$. We say a broken curve c is a *loop* at x if $c(0) = c(1) = x$. Given a loop l at x, parallel translation around l gives us a linear transformation of the tangent space X_x into itself. Notice that the "breaks" in the loop give us no difficulty since we can simply chain together the parallel translations.

Exercise 3.5: Show that the set of all those linear transformations given by parallel translation around loops at x forms a subgroup of the general linear group of the tangent space, $GL(X_x)$.

We denote this subgroup by $\Phi(X, x)$ and call it the *holonomy group of X at x.*

Actually $\Phi(X, x)$ depends also on the connection ∇, but it is usual to omit reference to the connection. In fact, we sometimes even omit the basepoint x and write $\Phi(X)$ (or just Φ) for $\Phi(X, x)$. To see some justification for this practice, note that $\Phi(X, x)$ has another definition which is similar to the definition of the fundamental group, $\pi_1(X, x)$. Namely, say two loops, l and l', are *holonomous* if parallel translation around l is the same as parallel translation around l'.

Exercise 3.6: Show that the relation of being "holonomous to" is an equivalence relation, and that the set of equivalence classes forms a group naturally isomorphic to $\Phi(X, x)$.

You may recall that $\pi_1(X, x)$ and $\pi_1(X, x')$ are conjugate (in something) and hence isomorphic. This is also the case with $\Phi(X, x)$ and $\Phi(X, x')$. In fact, if we pick a basis for X_x, and write the elements of $\Phi(X, x)$ as matrices with respect to this basis, $\Phi(X, x)$ becomes a subgroup of GL_n where n is the dimension of X. Then $\Phi(X, x)$ and $\Phi(X, x')$ are conjugate subgroups of GL_n.

Theorem 3.1 (Borel–Lichnerowicz): *Let X be a manifold with a connection, and let $\Phi_0(X)$ be the identity component of $\Phi(X)$. Then $\Phi_0(X)$ consists precisely of those holonomy classes of loops which are homotopic to the constant map: $[0,1] \to \{x\}$.*

For a proof of this theorem, see [67], page 33.

Corollary 3.1: *There is a surjective homomorphism $\eta : \pi_1(X) \to \Phi(X)/\Phi_0(X)$.*

Obviously you map a homotopy class, α, into the holonomy class of any loop in α. Let's look at an important example.

Example 3.1: Let π be a subgroup of A_n such that the orbit space $X = \mathbb{R}^n/\pi$ is an n-manifold, e.g. π could be a Bieberbach subgroup of $M_n \subset A_n$. Let $p : \mathbb{R}^n \to X$ be the projection map that sends each point to its orbit. Then it is easy to see that p is a local homeomorphism (every point in $\mathbb{R}^n$ has a neighborhood on which the restriction of p is a homeomorphism), and we define the differential structure on X so that p is locally a diffeomorphism. We define a connection ∇ on X as follows:

Let $x \in X$ and pick $r \in \mathbb{R}^n$ s.t. $p(r) = x$. Let $U_x \in X_x$. Since p has been made a local diffeomorphism, there is a unique vector $\tilde{U}_r \in (\mathbb{R}^n)_r$ with $dp_r(\tilde{U}_r) = U_x$. Similarly, if V is a vector field defined near x, there is a unique vector field $\tilde{V}$ defined near r with $dp(\tilde{V}) = V$. Put

$$\nabla_{U_x}(V) = dp_r(\tilde{\nabla}_{\tilde{U}_r}(\tilde{V}),$$

where $\tilde{\nabla}$ is the usual connection on $\mathbb{R}^n$.

If we define coordinates near x by picking a ball B around r small enough so that $p \mid B$ is injective (i.e. B contains no other points in the orbit of r), then $(p \mid B)^{-1}$ will be a coordinate map valid near x. Let's use $\tilde{x}_1, \ldots, \tilde{x}_n$ for the usual coordinates in $\mathbb{R}^n$ and $x_1, \ldots, x_n$ for these new coordinates in X. So $dp_r(\frac{\partial}{\partial \tilde{x}_i}(r)) = \frac{\partial}{\partial x_i}(x)$. Write $\tilde{U}_r = \sum u^i \frac{\partial}{\partial x_i}(r)$ $(u^i \in \mathbb{R})$ and $\tilde{V}_r = \sum v^i \frac{\partial}{\partial x_i}(r)$ where the v^i are functions defined near r. Then

$$\nabla_{U_x}(V) = \sum_k \left[\sum_i u^i \frac{\partial}{\partial \tilde{x}_i}(v^k) \right] \frac{\partial}{\partial x_k}(x). \tag{6}$$

Exercise 3.7: i) Prove (6).

ii) Show that the connection ∇ is independent of the choice of $r \in \mathbb{R}^n$ with $p(r) = x$. (**Hint:** If $p(r') = x$, then $r' = \alpha \cdot r$ for some $\alpha \in A_n$, and then use (6).)

iii) Show $p^*(\nabla) = \tilde{\nabla}$.

What can we say about the holonomy group of X? The idea is to use the last hint, and look at this picture.

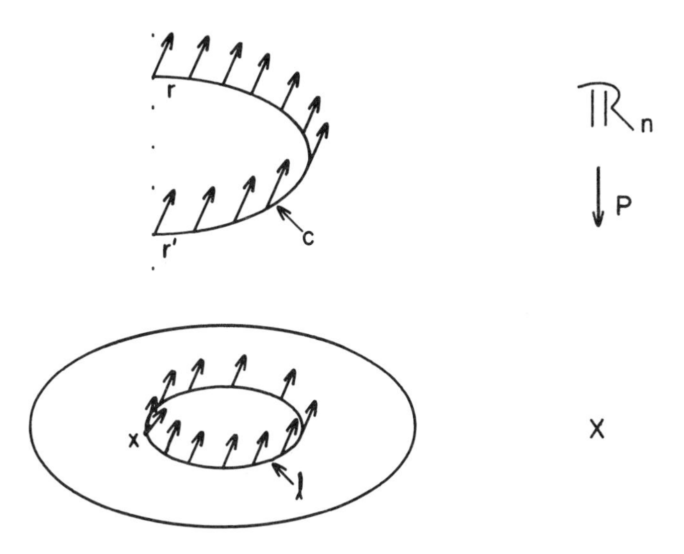

Say l is a loop at x. By the theory of covering spaces (see [58], page 151), there is a unique curve c in $\mathbb{R}^n$ with $c(0) = r$ and $p \circ c = l$, but c may not be closed (and won't be, unless l is homotopically trivial). Let $r' = c(1)$. Since $l(0) = l(1) = x, p(r') = p(r) = x$, so there is $(m, s) \in A_n$ with $(m, s) \cdot r = r'$. Now if $U_x \in X_x$, parallel translating U_x around l corresponds to moving $\tilde{U}_r$ parallel to itself around c, so the result $\|_0^1 (U_x)$ is just $dp_{r'}(\tilde{U}_{r'})$ where $\tilde{U}_{r'}$ is the vector at r' parallel to $\tilde{U}_r$ (remember, upstairs is $\mathbb{R}^n$). To see what the linear transformation induced by l on X_x is, we must compare dp_r with $dp_{p'}$, and this is merely the chain rule,

$$dp_{r'} = dp_r \circ d\alpha_r = dp_r \circ m$$

where α denotes the affine transformation of $\mathbb{R}^n$ defined by (m, s), so

$$\|_0^1 (U_x) = m \cdot U_x,$$

i.e. parallel translation around l is given by the linear transformation with matrix m. By drawing lines in $\mathbb{R}^n$ from r to each of the other points in its orbit and projecting to X, we get loops at x. Hence parallel translation around them is given by the elements of $r(\pi)$. We see that $\Phi(x)$ is isomorphic to $r(\pi)$, and this justifies the name "holonomy group" for $r(\pi)$ when π is a Bieberbach group.

Now covering space theory tells us that $\mathbb{R}^n$ is the universal covering space of X, π is the group of deck transformation, and the fundamental group $\pi_1(X)$ is isomorphic to π. Also since π is a Bieberbach subgroup of M_n, Bieberbach's First Theorem tells us that $\Phi(X)$ (which is $r(\pi)$), is finite, so $\Phi_0(X)$ is trivial (i.e. consists of just the identity element). Hence the map $\eta : \pi_1(X) \to \Phi(X)/\Phi_0(X)$ defined in Corollary 3.1 is precisely the rotational part map $r : \pi \to r(\pi)$.

Our next collection of definitions and remarks serves to give some special geometric significance in the case that π is a Bieberbach group.

Definition 3.5: Let X be an n-manifold and U and V vector fields on X. Then $[U,V]$ is the vector field defined by

$$[U,V](f) = U(V(f)) - V(U(f)), \quad \forall f \in C^\infty(X). \tag{7}$$

Exercise 3.8: i) Show that (7) does indeed define a vector field, although $f \mapsto U(V(f))$ is not a vector field.

ii) Suppose we have a coordinate system $x_1, \ldots, x_n$, and $U = \sum u^i \frac{\partial}{\partial x_i}$ and $V = \sum v^i \frac{\partial}{\partial x_i}$ where the coordinate system is defined. Show that

$$[U,V] = \sum_{k,i} \left(u^k \frac{\partial}{\partial x_k}(v^i) - v^k \frac{\partial}{\partial x_k}(u^i) \right) \frac{\partial}{\partial x_i}.$$

iii) Let $x \in X$ and let c_1 be a curve in X with $c_1(0) = x$ and $(\dot{c}_1)_0 = U_x$. Let $\epsilon > 0$, and take a curve c_2 with $c_2(0) = c_1(\epsilon)$ and $(\dot{c}_2)_0 = V_{c_1(\epsilon)}$. Let c_3 be a curve s.t. $c_3(0) = c_2(\epsilon)$ and $(\dot{c}_3)_0 = -U_{c_2(\epsilon)}$.

Finally let c_4 be a curve s.t. $c_4(0) = c_3(\epsilon)$ and $(\dot{c}_4)_0 = -V_{c_3(\epsilon)}$. There is no reason to suppose this little "rectangle" is closed, so let c be defined by $c(\epsilon^2) = c_4(\epsilon)$. Show that $[U, V]_x = (\dot{c})_0$ and thus is independent of the choice of c_1, c_2, c_3, c_4 and c. (More precisely, show $\lim_{t \to +0}(\dot{c})_t = [U, V]_x$.) (There is a marvellous discussion of the procedure on pages 33-34 of [66].)

$[U, V]$ is called the *bracket* or *Lie bracket* of U and V. Note that it does not depend on a connection and does not work for tangent vectors, i.e. U and V must be vector *fields*.

Definition 3.6: Let U and V be vector fields on a manifold X with connection ∇. Define a transformation $R(U, V)$ of vector fields to vector fields by

$$R(U, V) \cdot W = -\nabla_U(\nabla_V W) - \nabla_V(\nabla_U W) + \nabla_{[U,V]} W \tag{8}$$

for any vector field W. We sometimes write symbolically

$$R(U, V) = \nabla_{[U,V]} - [\nabla_U, \nabla_V].$$

R is called the *curvature* or *curvature mapping* of X (with ∇).

Lemma 3.3: *Let $x \in X$. Then $[R(U, V) \cdot W]_x$ depends only on U_x, V_x , and W_x. If U_x and V_x are tangent vectors at x, $R(U_x, V_x)$ is a map $:X_x \to X_x$. The mapping*

$$(U_x, V_x, W_x) \mapsto R(U_x, V_x) \cdot W_x$$

is trilinear.

For a proof, see page 51 of [61].

This lemma shows that we can consider R to be a map:$X_x \times X_x \to \mathrm{Hom}(X_x, X_x)$.

Exercise 3.9: i) Show that for $\mathbb{R}^n$ with the usual connection, R is the 0 map. (This is easy.)

ii) Let $S_2 = \{r \in \mathbb{R}^3 : \|r\| = 1\}$, i.e. S_2 is the ordinary two dimensional sphere (and it's even in the usual place). Let $i : S_2 \hookrightarrow \mathbb{R}^3$, and let $\nabla = i^*(\text{usual connection on } \mathbb{R}^3)$. Show that this connection has non-zero curvature. (This is harder.)

Remark: Exercise 3.9 shows that, at least in the simplest cases, curvature is what we think it should be. In the general setting of connections, it is difficult to give an equally satisfying geometric interpretation. By looking at the "rectangle" of Exercise 3.8, part iii), we can see a pretty geometric interpretation, but it still gives no particular reason why R should be called "curvature".

Exercise 3.10: Let $W_x \in X_x$. Parallel translate W_x around the "rectangle," i.e. start at $c_1(0)$ then go to $c_1(\epsilon)$, then go to $c_2(\epsilon)$, then go to $c_3(\epsilon)$, then go to $c_4(\epsilon)$, and finally from $c_4(\epsilon)$ back to x along any c. Show that the tangent vector at x you get is

$$W_x + \epsilon^2 R(U_x, V_x) \cdot W_x + \text{terms of high order in } \epsilon.$$

(This is harder yet.)

This result gives rise to a relation between curvature and holonomy which culminates in the next theorem. To state this theorem, recall that the holonomy group $\Phi(X, x)$ can be thought of as a subgroup of GL_n by choosing a basis for X_x. We need to talk about a map exp from M_n, the algebra of all $n \times n$ real matrices, to GL_n.

Definition 3.7: The *exponential map*, $\exp : M_n \to GL_n$, is defined by the formula

$$\exp A = I + A + A^2/2! + A^3/3! + \ldots = \sum_{i=0}^{\infty} \frac{A^i}{i!}. \tag{9}$$

The basic properties of this map are developed in the following exercises. Answers to these exercises can be found in Chapter I of [24].

Exercise 3.11: i) Show that the series in (9) always converges.

ii) Show $\det(\exp A) = \exp(\det A)$, so $\det \exp A \neq 0, \forall A \in M_n$.

iii) Show $\exp A$ is in O_n iff A is skew symmetric.

Definition 3.8: A subspace N of M_n (considered as a real vector space) is called a *Lie subalgebra* of M_n if $A, B \in N \Rightarrow AB - BA \in N$. We write $[A, B] = AB - BA$ as in Definition 3.5 and call it the *Lie bracket* of A and B.

Exercise 3.12: If S is any subset of M_n, show there is a unique smallest Lie subalgebra of M_n containing S. This is called the *Lie algebra generated by S*.

Now if $x \in X$ and U_x and $V_x \in X_x$, then $R(U_x, V_x)$ is a linear map:$X_x \to X_x$. If we have chosen a basis b for X_x, we can consider $R(U_x, V_x)$ as a member of M_n, say $R(U_x, V_x)_b$.

Theorem 3.2 (Ambrose–Singer Holonomy Theorem): *Let X be a manifold with connection ∇. Let $x \in X$ and suppose we have chosen a basis b_0 for X_x. Suppose that y is in the same component of X as x. Let N be the Lie subalgebra of M_n generated by*

$$\{R(U_y, V_y)_b : b \text{ is a parallel translate of } b_0 \text{ to } y \text{ and } U_y, V_y \in X_y\}.$$

Then $\exp(N) = \Phi(X, x)$.

This remarkable theorem was first stated in pretty much the above form by E. Cartan (see [15]), but his version of the proof was somewhat vague (to say the least). Ambrose and Singer gave the first understandable proof in a more general setting in [1]. Subsequently, Nomizu gave a different proof (see page 39 of [67]) which is the one usually referred to now.

Corollary 3.2: *R is identically zero iff $\Phi(X)$ is totally disconnected, i.e. $\Phi_0(X) = \{I\}$. Hence in this case, the map η maps: $\pi_1(X) \xrightarrow{\eta} \Phi(X) \longrightarrow 0$.*

Definition 3.9: If R is identically zero, we say the connection is *flat*.

Remark: We are interested in a special kind of flat connection to be defined in the next section. However, there is an interpretation of "curvature zero" we would like to mention here. Locally (i.e. in a neighborhood of each point) an n-manifold X looks topologically (even differentiably) just like $\mathbb{R}^n$, but when we put a connection ∇ on X, ∇ may not look anything like the usual connection on $\mathbb{R}^n$. If we want to be inelegant and write everything in coordinates (and we can since we are working only locally), then we know that ∇ is determined by n^3 functions $\Gamma^k_{i,j}(i, j, k = 1, 2, \dots n)$, and the usual connection on $\mathbb{R}^n$ has $\Gamma^k_{i,j} = 0$. So we would like to know conditions on ∇ near $x \in X$ so that we can introduce coordinates such that $\Gamma^k_{i,j}$ will be zero for all i, j, k in these coordinates. If you write everything out, you get a set of partial differential equations. The general theory of such PDE's tells you that a solution exists iff a certain expression vanishes. The situation is a generalization of the equations $\frac{\partial f}{\partial x} = P$ and $\frac{\partial f}{\partial y} = Q$. Since

$$\frac{\partial^2 f}{\partial x \partial y} = \frac{\partial^2 f}{\partial y \partial x}$$

(recall everything is infinitely differentiable), we must have

$$\frac{\partial P}{\partial y} - \frac{\partial Q}{\partial x} = 0,$$

and if we have this, we can find f. In the case at hand, the expression that must vanish in order to get coordinates with $\Gamma^k_{i,j} = 0$ is precisely R. In coordinates, R looks like $R^l_{i,j,k}$, i.e.

$$R(\frac{\partial}{\partial x_i}, \frac{\partial}{\partial x_j}) \cdot \frac{\partial}{\partial x_k} = \sum R^l_{i,j,k} \frac{\partial}{\partial x_l}.$$

There is a discussion of this on pages 301-303 in [76].

Recall the example mentioned earlier in this section. $X = \mathbb{R}^n/\pi$ where $\pi \subset A_n$. In this case we had $R \equiv 0$, and in the case ∇ is "nice," these are the only flat examples. We now explain "nice."

Definition 3.10: Let U and V be vector fields on a manifold X with connection ∇. Define a new vector field $T(U, V)$ by

$$T(U, V) = \nabla_U V - \nabla_V U - [U, V]. \tag{10}$$

Exercise 3.13: Let $x \in X$. Show that $T(U, V)_x$ depends only on U_x and V_x, so T can be considered to be a map : $X_x \times X_x \to X_x$. Show this map is bilinear.

T is called the *torsion* or *torsion mapping* of ∇.

Exercise 3.14: Let $x_1, \ldots, x_n$ be coordinates at x, and write

$$T(\frac{\partial}{\partial x_i}, \frac{\partial}{\partial x_j}) = \sum T^k_{i,j} \frac{\partial}{\partial x_k}.$$

Show

$$T^k_{i,j} = \Gamma^k_{i,j} - \Gamma^k_{j,i}.$$

If $T \equiv 0$, we say ∇ is a *symmetric* or *torsionfree* connection.

The example mentioned earlier had $T \equiv 0$. Part of being "nice" is being symmetric. An excellent discussion of the geometric meaning of torsion can be found on page 75 of [66].

Definition 3.11: A curve c on X is called a *geodesic* if $\nabla_{\dot{c}}\dot{c} = 0$.

Exercise 3.15: Let $x \in X$ and $U_x \in X_x$. Show there is a unique geodesic g on X (defined in $[0, \epsilon)$ for some $\epsilon > 0$) with $g(0) = x$ and $\dot{g}_0 = U_x$. (**Hint:** Pick coordinates and use the fundamental existence and uniqueness theorem of differential equations.)

The main interest in geodesics is that they are the "straight lines" of "curved" manifolds and that in the riemannian case, they locally minimize distance (see Theorem 4.1). This property is not of much interest to us, but if it is to you, look on pages 59-66 of [61].

Now we would like to define a map called, strangely enough, the *exponential map* which goes, strangely enough, from X_x to X. One way to think of what we want to do is do "wrap the tangent space around the manifold" (J. Eells). The last exercise should give us the following idea: Set $\exp(U_x) = g(1)$ where g is the unique geodesic with $g(0) = x$ and $\dot{g}_0 = U_x$. The difficulty is that we don't know that $g(1)$ is defined. It turns out that if U_x is small enough, g can be extended to all of $[0, 1]$ so $g(1)$ is defined. That is, there is always a neighborhood of 0 in X_x where exp can be defined in this manner. For some especially "good" manifolds with especially "good" connections, exp can be defined on all of X_x.

Definition 3.12: Suppose X is a manifold with connection ∇ and suppose that $\forall x \in X$ and $\forall U_x \in X_x$, the unique geodesic g with $g(0) = x$ and $\dot{g}_0 = U_x$ can be defined on all of $[0, 1]$. Then we say X (with ∇) is *complete*, or we say that ∇ is *complete.*

Exercise 3.16: Show that the above example $X = \mathbb{R}^n/\pi$ is complete. Show that $\mathbb{R}^n - \{0\}$ is not complete.

Now we can say what "nice" is. A connection is "nice" if it is both symmetric and complete. We have the following theorem alluded to above:

Theorem 3.3 : *Let X be a simply connected and connected n-manifold with a complete, symmetric, flat connection. Then X is affinely equivalent to $\mathbb{R}^n$ with the usual connection.*

Corollary 3.3: *Let X be a connected n-manifold with a complete, symmetric, flat connection. Then there is a subgroup π of A_n s.t. X is affinely equivalent to $\mathbb{R}^n/\pi$.*

The corollary follows easily from the theorem by using the theory of covering spaces. We will not prove the theorem here (see [49] page 211). The proof depends on showing that X is an abelian Lie group. We will prove, in the next section, a somewhat stronger theorem in a special case which holds more interest for us.

4. Riemannian Structures

The most interesting connections arise from the following construction:

Definition 4.1: Let X be a manifold. A *riemannian structure* on X is a map which assigns to each $x \in X$ a positive definite inner product $\langle\,,\rangle_x$ on X_x s.t. if U and V are (smooth) vector fields on X, the function $x \mapsto \langle U_x, V_x\rangle_x$ is a smooth function. We say X (with $\langle\,,\rangle$) is a *riemannian manifold.*

Exercise 4.1: Show that every manifold has some riemannian structure.

This exercise is non-trivial. You need to know about "partitions of unity" which is a device that will allow you to define a riemannian structure separately in each coordinate system and then "fit them together." If we have coordinates $x_1, \ldots, x_n$, then we define functions

$$g_{i,j} = \langle \frac{\partial}{\partial x_i}, \frac{\partial}{\partial x_j} \rangle_x$$

which will determine the riemannian structure in the coordinate system. Conversely, if $(g_{i,j}(x))$ is any symmetric positive definite matrix for each x, and it is smooth as a function of x, then we can use it to define a riemannian structure where the coordinates are valid. $\mathbb{R}^n$ has a natural riemannian structure given by

$$(g_{i,j}(x)) = I \quad \forall x \in \mathbb{R}^n.$$

Definition 4.2: We use the word *isometry* to describe the following three different notions:

i) A linear transformation between inner product spaces that preserves the inner product, i.e. $f : V \to W$ s.t.

$$\langle f(v_1), f(v_2)\rangle_W = \langle v_1, v_2\rangle_V.$$

ii) A diffeomorphism $f : X \to Y$ between riemannian manifolds s.t. $df_x : X_x \to Y_{f(x)}$ is an isometry in the sense of i).

iii) A map between metric spaces that preserves distance.

Proposition 4.1: *Let X be a riemannian manifold. Then there is a unique symmetric connection on X s.t. parallel translation is an isometry (in the sense of i) above), i.e. if c is a curve on X, then $\|_0^1 : X_{c(0)} \to X_{c(1)}$ is an isometry.*

A proof (which is not difficult) can be found on page 48 of [61]. This connection is called the *riemannian* or *Levi–Civita connection.*

Before proceding to our main interest (flat manifolds), we will record here some definitions and theorems to give you some of the flavor of riemannian geometry.

Definition 4.3: Let X be a riemannaian manifold and c a curve on X. Define the *length* of c by

$$l(c) = \int_0^1 \langle \dot{c}_t, \dot{c}_t \rangle_{c(t)}^{1/2} dt.$$

For $x, y \in X$ define the *distance from x to y* by

$$d(x,y) = \text{glb}\{l(c) : c(0) = x \text{ and } c(1) = y\}.$$

Proposition 4.2: *X together with d is a metric space whose metric topology coincides with the given topology on X.*

For a proof see [61], pages 59 and 62.

Definition 4.4: A geodesic g from x to y is said to be *minimal* if $l(g) \leq l(c)$ for any curve c from x to y.

Theorem 4.1 (Hopf–Rinow): *The riemannian connection of X is complete iff either*

i) X together with d is a complete metric space, or

ii) there is a minimal geodesic between any two points of X.

See page 62 of [61] for a discussion of this result.

Corollary 4.1: *Any compact riemannian manfold is complete.*

Proof: A compact metric space is complete. ☐

Theorem 4.2 : *Let X and Y be riemannian manifolds. Then a map $f : X \to Y$ is an isometry in the sense of ii) iff it is an isometry in the sense of iii).*

5. Flat Manifolds

We finally get to the object of principal interest in this chapter, namely flat manifolds, i.e. riemannian manifolds whose riemannian connection has identically zero curvature. We usually call them simply "flat manifolds." From here on we assume all manifolds are connected.

First we consider the local situation in which curvature vanishes in a neighborhood of a point. The basic lemma is the following:

Lemma 5.1: *Let X be a riemannian manifold and $x_0 \in X$. Suppose the curvature $R(U_x, V_x)$ is the zero map $\forall U_x, V_x \in X_x$ and $\forall x$ near x_0. Then given any $U_{x_0} \in X_{x_0}$, there is a vector field U defined near x_0 whose value at x_0 is U_{x_0} and which is parallel, i.e. $\nabla_V U = 0$ near x_0 for any vector field V.*

Proof: Let $x_1, \ldots, x_n$ be any coordinate system valid near x_0. Let c_1 be the coordinate curve corresponding to x_1, i.e. $x_1(c_1(t)) = t$ and $x_i(c_1(t)) = 0$ for $i > 1$. Define U along c_1 by parallel translation of U_{x_0} along c_1.

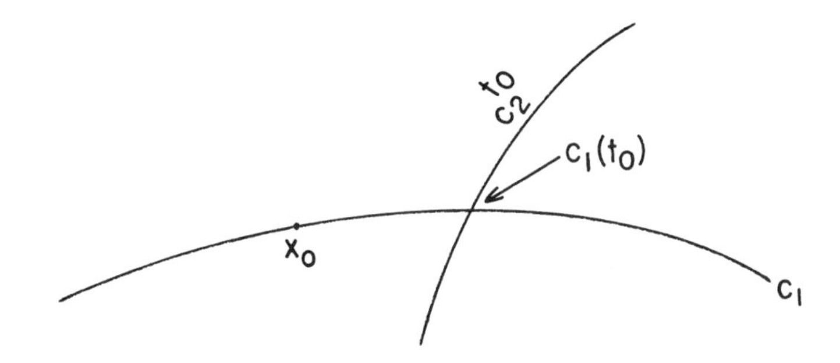

Let $t_0 \in [0,1]$ and define $c_2^{t_0}$ by

$$x_1(c_2^{t_0}(t)) = t_0,\ x_2(c_2^{t_0}(t)) = t,\ \text{and } x_i(c_2^t) = 0, \text{ for } i > 2.$$

We define U along $c_2^{t_0}$ by parallel translating $U_{c_1(t_0)}$. Since we can do this for each $t_0 \in [0,1]$, we have defined U on the two-dimensional surface spanned by x_1 and x_2. Clearly $\nabla_{\frac{\partial}{\partial x_2}} U = 0$ on this surface. Now

$\left[\frac{\partial}{\partial x_1}, \frac{\partial}{\partial x_2}\right] = 0$, so by the definition of curvature, equation (8),

$$-\nabla_{\frac{\partial}{\partial x_1}}(\nabla_{\frac{\partial}{\partial x_2}} U) - \nabla_{\frac{\partial}{\partial x_2}}(\nabla_{\frac{\partial}{\partial x_1}} U) = R(\frac{\partial}{\partial x_1}, \frac{\partial}{\partial x_2}) \cdot U = 0 \qquad (11)$$

since R is zero near x_0. Since $\nabla_{\frac{\partial}{\partial x_2}} U = 0$, $-\nabla_{\frac{\partial}{\partial x_2}}(\nabla_{\frac{\partial}{\partial x_1}} U) = 0$, i.e. $\nabla_{\frac{\partial}{\partial x_1}} U$ is parallel along $c_2^{t_0}$. But U is clearly parallel along c_1, so $\left(\nabla_{\frac{\partial}{\partial x_1}} U\right)_{c_1(t_0)} = 0$, so $\nabla_{\frac{\partial}{\partial x_1}} U$ is the parallel transport of the zero vector, and hence $\nabla_{\frac{\partial}{\partial x_1}} U = 0$ on the surface also.

Now continue the process. Let $c_3^{t_0,t_1}$ be defined by

$$\begin{aligned} x_1\left(c_3^{t_0,t_1}(t)\right) &= t_0, \\ x_2\left(c_3^{t_0,t_1}(t)\right) &= t_1, \\ x_3\left(c_3^{t_0,t_1}(t)\right) &= t, \text{ and} \\ x_i\left(c_3^{t_0,t_1}(t)\right) &= 0, \text{ for } i > 3. \end{aligned}$$

Define U along $c_3^{t_0,t_1}$ by parallel translation of $U_{c_2^{t_0}(t_1)}$ along $c_3^{t_0,t_1}$. So $\nabla_{\frac{\partial}{\partial x_3}} U = 0$ in the three-dimensional subset of X spanned by x_1, x_2, and x_3. Then two equations like (11) will show $\nabla_{\frac{\partial}{\partial x_1}} U = \nabla_{\frac{\partial}{\partial x_2}} U = 0$, and so on.

Now we have a vector field U defined near x_0 with $\nabla_{\frac{\partial}{\partial x_i}} U = 0$ for $i = 1, 2, \ldots, n$. If V is any vector field defined near x_0, we can write $V = \sum_i V_i \frac{\partial}{\partial x_i}$, so $\nabla_V U = \sum_i V_i \nabla_{\frac{\partial}{\partial x_i}} U = 0$. □

Compare the above proof with the "little rectangle" of Exercise 3.8, part iii). It is not hard from this lemma to show that parallel translation is independent of path near x_0, i.e. where curvature vanishes. Note that the fact X was riemannian played no part in the proof; this is really a lemma about a manifold with a connection whose curvature vanishes in an open set. This lemma is the local part of the proof of Theorem 5.2.

Proposition 5.1: *Let X be a riemannian manifold and $x_0 \in X$. Suppose the curvature $R(U_x, V_x)$ is the zero map $\forall U_x, V_x \in X_x$ and $\forall x$ near x_0. Then there are coordinates $y_1, \ldots, y_n$ valid near x_0 s.t.*

$$g_{i,j}(x) = \langle \frac{\partial}{\partial y_i}(x), \frac{\partial}{\partial y_j}(x) \rangle_x = \delta_{i,j}$$

for all x near x_0.

Proof: By the Gram–Schmidt process, we can pick an orthonormal basis $U^1_{x_0}, \dots, U^n_{x_0}$ for X_{x_0}, i.e.

$$\langle U^i_{x_0}, U^j_{x_0} \rangle_{x_0} = \delta_{i,j}.$$

By Lemma 5.1, we get parallel vector fields $U^1, \dots, U^n$ near x_0, and since parallel translation is an isometry,

$$\langle U^i_x, U^j_x \rangle_x = \delta_{i,j}.$$

for x near x_0.

We want to find a coordinate sysytem $y_1, \dots, y_n$ near x_0 s.t. $\frac{\partial}{\partial y_i} = U^i$. There is a famous theorem of Frobenius which is just what we want.

Theorem 5.1 (Frobenius): *Let X be a manifold and $U^1, \dots, U^n$ vector fields defined near $x_0 \in X$. Suppose*

i)$U^1_x, \dots, U^n_x$ spans X_x for all x near x_0, and

ii) $[U^i, U^j] = 0$.

Then there exists a coordinate sysytem $y_1, \dots, y_n$ valid near x_0 s.t. $\frac{\partial}{\partial y_i} = U^i$.

A proof can be found, for example, on page 89 of [24].

So we must show our $U^1, \dots, U^n$ satisfy i) and ii). i) is immediate since parallel translation is an isometry and hence a vector space isomorphism. For ii), recall that the riemannian connection is symmetric, i.e. $T = 0$. So by (10),

$$[U^i, U^j] = \nabla_{U^i} U^j - \nabla_{U^j} U^i.$$

But since both U^i and U^j are parallel, $\nabla_{U^i} U^j = \nabla_{U^j} U^i = 0$, and ii) is satisfied. □

Remark: In the remarks after Definition 3.9, we gave an interpretation of curvature as the obstruction to parallel translation being independent of path. This interpretation could provide an alternate proof of Lemma 5.1. The above proposition shows that in the riemannian case, curvature can be regarded as the obstruction to euclidean geometry being locally valid. The reference to the "general theory of PDE's" given in those remarks is exactly the Frobenius theorem.

The global version of Proposition 5.1 is one third of the Clifford–Klein Theorem first proved by H. Hopf ([44]). The other two thirds tell what simply connected complete riemannian manifolds with constant positive or constant negative curvature look like.

Theorem 5.2 (One Third of Clifford–Klein): *Let X be a (connected) simply connected, complete flat (riemannian) n-manifold. Then X is isometric to $\mathbb{R}^n$ (with the usual riemannian structure).*

Proof: Let $x_0 \in X$ and let $U^1_{x_0}, \ldots, U^n_{x_0}$ be an orthonormal basis for X_{x_0}. Let $\bar{x} \in \mathbb{R}^n$ be arbitrary, and $\bar{g}$ be the line in $\mathbb{R}^n$ with $\bar{g}(0) = 0$ and $\bar{g}(1) = \bar{x}$. Let $e_1, \ldots, e_n$ be the usual basis of $\mathbb{R}^n$ (or $(\mathbb{R}^n)_0$). Write $\dot{\bar{g}}_0 = \sum G_i e_i$ and let $U_{x_0} \in X_{x_0}$ be defined by

$$U_{x_0} = \sum G_i U^i_{x_0}.$$

Define $F : \mathbb{R}^n \to X$ by $F(\bar{x}) = x$ where $x = g(1)$ and g is the unique geodesic with $g(0) = x_0$ and $\dot{g}_0 = U_{x_0}$. (F is really the exponential map.) Since X is complete (and connected), F is well-defined and surjective.

Now it follows from Proposition 5.1 that if $\bar{x}$ is near 0 in $\mathbb{R}^n$, we can introduce coordinates $x_1, \ldots, x_n$ near x $(= F(\bar{x}))$ s.t. if $U_x = \sum A_i U^i_{x_0}$ and $V_x = \sum B_i U^i_{x_0}$, then

$$\langle U_x, V_x \rangle_x = \sum A_i B_i.$$

Exercise 5.1: Show this implies that F is an isometry near $\bar{x}$.

Let $S = \{\bar{x} \in \mathbb{R}^n : F \text{ is an isometry near } \bar{x}\}$. Clearly S is open. We want to show S is closed.

Suppose there is $\bar{y} \in \mathbb{R}^n$ with the property that if g is the line from 0 to y, F is an isometry near each point of $g([0,1])$. Let $y = F(\bar{y})$. Since $R \equiv 0$, there is a small neighborhood N of y which is isometric to a small neighborhood $\bar{N}$ of y, but possibly not by F. Let $\bar{z} \in g([0,1]))$ be close enough to $\bar{y}$ so that $\bar{z}$ has a neighborhood $\bar{M}$ with $\bar{M} \subset \bar{N}$ and $F \mid \bar{M}$ is an isometry.

Put $z = F(\bar{z})$ and $M = F(\bar{M})$. Let $H : \bar{N} \to N$ be any isometry. Then

$$(H^{-1} \mid M) \circ (F \mid \bar{M}) : \bar{M} \to \bar{M}$$

is (can be extended to) an isometry of $\mathbb{R}^n$, i.e. a rigid motion. Call it α and let $H_1 = H \circ \alpha$. Then $H_1 \mid \bar{M} = H \circ \alpha \mid \bar{M} = H \circ H^{-1} \circ F \mid \bar{M} = F \mid \bar{M}$, and H_1 is still an isometry:$\bar{N} \to N$.

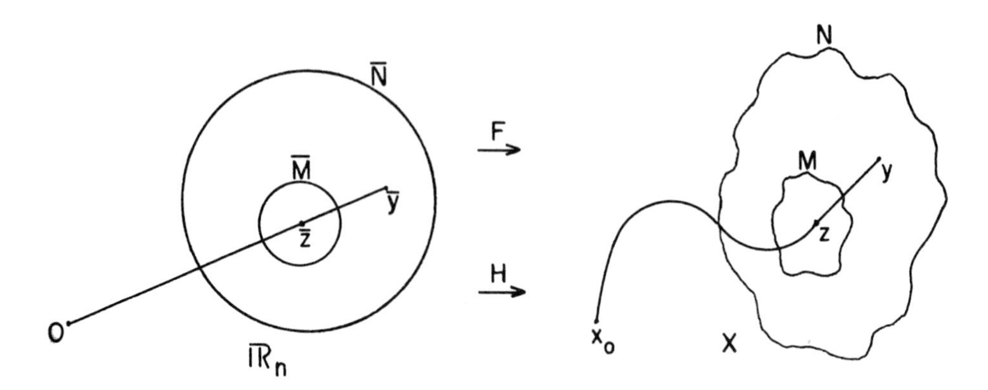

Now H_1 and F are both isometries on $\bar{N}$, and they agree on $\bar{M}$, so they must agree on the cone from 0 through the points of $\bar{M}$, and this cone contains a neighborhood of $\bar{y}$.

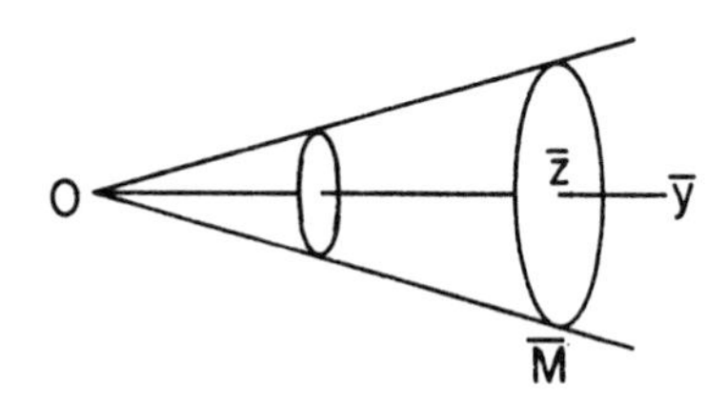

Since H_1 is an isometry on a neighborhood of $\bar{y}$ by construction, F is a isometry in a neighborhood of $\bar{y}$, and $\bar{y} \in S$, so S is closed. Hence every point $\bar{x} \in \mathbb{R}^n$ has a neighborhood on which F is an isometry.

Now we show that F is a covering map. We must show that every point $x \in X$ has a neighborhood N s.t. $F^{-1}(N)$ is a disjoint union of open sets $\bar{N}_1, \ldots, \bar{N}_r$ in $\mathbb{R}^n$ for which $F \mid N_i$ is a homeomorphism for each i. Suppose not. Then $\forall \epsilon > 0$, we could find points $\bar{x}_1, \bar{x}_2 \in \mathbb{R}^n$ s.t. $F(\bar{x}_1) = F(\bar{x}_2) = x_0$ and $|\bar{x}_1 - \bar{x}_2| < \epsilon$. Let $\bar{g}$ be the line from $\bar{x}_1$ to $\bar{x}_2$ and let $g = F(\bar{g})$. Thus $l(\bar{g}) < \epsilon$. Hence $l(g) < \epsilon$, and g is a closed geodesic at x_0. This would say that there are closed geodesics of arbitrarily small length which is absurd since Proposition 5.1 says X is locally $\mathbb{R}^n$.

Therefore, X is covered by $\mathbb{R}^n$, but since X is simply connected, F must be a homeomorphism, and since it is a local isometry, it must be an isometry. □

Corollary 5.1: *Let X be a (connected) complete flat (riemannian) n-manifold. Then there is a discrete torsionfree subgroup, π, of $\mathcal{M}_n$ such that*

X is isometric to $\mathbb{R}^n/\pi$. Furthermore, if X is compact, π is a Bieberbach subgroup.

Exercise 5.2: Prove this corollary.

Now we can state the three theorems of Bieberbach in the context of flat manifolds.

Theorem 5.3 (Bieberbach's First): *Let X be a compact flat manifold. Then X is covered by a flat (riemannian) torus, and the covering map is a local isometry. Furthermore, the holonomy group $\Phi(X)$ is finite.*

Proof: $X = \mathbb{R}^n/\pi$ is covered by $\mathbb{R}^n/\pi \cap \mathbb{R}^n$ which by Theorem 3.1 is a torus. Furthermore, $\Phi(X)$ is isomorphic to $r(\pi)$ (see Example 3.1) which is finite also by Theorem 3.1. □

Theorem 5.4 (Bieberbach's Second): *Let X and Y be compact flat manifolds and suppose that $\pi_1(X)$ is isomorphic to $\pi_1(Y)$. Then X and Y are affinely equivalent.*

Proof: We consider $\pi_1(X)$ and $\pi_1(Y)$ as Bieberbach subgroups of $\mathcal{M}_n$. By Theorem 4.1, there is $\alpha \in \mathbb{R}^n$ s.t. if $F : \pi_1(X) \to \pi_i(Y)$ is an isomorphism, $F(\beta) = \alpha\beta\alpha^{-1} \quad \forall \beta \in \pi_1(X)$. Let

$$p_X : \mathbb{R}^n \to X = \mathbb{R}^n/\pi_1(X) \text{ and } p_Y : \mathbb{R}^n \to Y = \mathbb{R}^n/\pi_1(Y)$$

be the projections (or covering maps). Define $f : X \to Y$ by

$$f(x) = p_Y \circ \alpha \circ p_X^{-1}(x) \text{ for } x \in X.$$

To see that this is well-defined, let $\bar{x} \in p_X^{-1}(x)$ and $\beta \in \pi_1(X)$. We must show $p_Y \circ \alpha(\beta \cdot \bar{x}) = p_Y \circ \alpha(\bar{x})$. But $\alpha\beta\alpha^{-1} = \gamma \in \pi_1(Y)$, so $\alpha\beta = \gamma\alpha$ and

$$p_Y(\alpha\beta \cdot \bar{x}) = p_Y(\gamma\alpha \cdot \bar{x}) = p_Y(\alpha \cdot \bar{x}),$$

so f is well-defined.

Exercise 5.3: Show f is an affine equivalence.

This completes the proof. □

Theorem 5.5 (Bieberbach's Third): *There are only finitely many affine equivalence classes of compact flat manifolds in any dimension.*

The proof is immediate from the previous theorems and Theorem 7.1 of Chapter I.

6. Conjectures and Counterexamples

A. *Locally Affine Spaces*

In the Second and Third Theorems above, one puts in riemannian information but gets out only affine results, e.g. in the Second Theorem, X and Y are not isometric, only affinely equivalent. Now you wouldn't expect to get an isometry since a big fat torus certainly has the same fundamental group as a little skinny one, but you might hope to put in only affine information to get the affine results. After all, if you only have a manifold X with connection ∇, you know what it means for ∇ to be flat. You had better add the hypothesis of completeness since an arbitrary connection on a compact manifold need not be complete. If you make the connection symmetric for good measure, there is a name for this kind of manifold.

Definition 6.1: A manifold X with a complete, symmetric flat connection is called a *locally affine space.*

This terminology is somewhat unfortunate. The idea, of course, is that X looks locally like $\mathbb{R}^n$, at least affinely. The unfortunate part arises when some writers extend this reasoning to flat (riemannian) manifolds, and call them *locally Euclidean spaces.* They mean that locally these spaces are like $\mathbb{R}^n$ *geometrically*, but the term "locally euclidean" has already been used for spaces that are locally like $\mathbb{R}^n$ *topologically*, i.e. "manifolds" which may not be Hausdorff.

In any case, you might wonder if a locally affine space X would satisfy the results of Bieberbach's three theorems. After all, you have Theorem 3.3, which says X is affinely equivalent to $\mathbb{R}^n/\pi$ where π is a subgroup of A_n, so we can at least get started.

The first problem arises because of Open Problem 1.1 of Chapter I. We do not know which subgroups π of A_n have orbit spaces that are n-manifolds. We know they should be discrete and torsionfree, but that's certainly not enough since we have seen that torsionfree doesn't guarantee a free action. Even if we look at discrete subgroups which act freely, we still don't know if their orbit spaces are n-manifolds. The situation doesn't improve if we throw in the hypothesis that π is uniform, i.e. $\mathbb{R}^n/\pi$ is compact. We can try to finesse the problem by studying discrete uniform subgroups of A_n and trying to prove Bieberbach type theorems about them. Then we will obtain results for compact locally affine spaces, since the fundamental groups of such spaces are certainly contained in this class of subgroups of A_n.

The next problem is that the analogue of Bieberbach's First Theorem is totally false. There are two extremely interesting counterexamples which we have already given. Recall Example 2.7. Topologically, X is merely the two-dimensional torus, but with a very strange connection. $\pi \in A_n$ has no pure translations at all, and $r(\pi)$ is $\mathbb{Z} \oplus \mathbb{Z}$ which is hardly finite. In fact, $r(\pi)$ is not even a closed subgroup of A_2.

While this example appears to demolish any hopes for the First Theorem for subgroups of A_n, one might argue that what we really want is a free abelian subgroup of π of rank n. In the case of rigid motions, this is $\pi \cap \mathbb{R}^n$, but perhaps in the case of affine motions, it is just some subgroup of π. In the above example, π itself is free abelian of rank 2. Example 2.6 destroys this hope. It is a subgroup of A_3 with no free abelian subgroups of rank 3 at all.

L. Auslander has also given some more examples which show the Third Theorem is also hopeless in the locally affine case (see [3]). Consider the following elements of A_3:

$$K(i) = \begin{pmatrix} 1 & 0 & 0 & 1/i \\ 0 & 1 & 0 & 0 \\ 0 & 0 & 1 & 0 \\ 0 & 0 & 0 & 1 \end{pmatrix},$$

$$L = \begin{pmatrix} 1 & 0 & 0 & 0 \\ 0 & 1 & 0 & 1 \\ 0 & 0 & 1 & 0 \\ 0 & 0 & 0 & 1 \end{pmatrix}, \text{ and}$$

$$J = \begin{pmatrix} 1 & 1 & 0 & 0 \\ 0 & 1 & 0 & 0 \\ 0 & 0 & 1 & 1 \\ 0 & 0 & 0 & 1 \end{pmatrix}.$$

Let $\pi(i)$ be the subgroup of A_3 generated by these elements.

Exercise 6.1: Show that $X(i) = \mathbb{R}^3/\pi(i)$ is a 3-manifold. Further show

$$\pi(i)/[\pi(i), \pi(i)] \approx \mathbb{Z} \oplus \mathbb{Z} \oplus \mathbb{Z}_i$$

(which is $H_1(X;\mathbb{Z})$).

Hence no two of the groups $\pi(i)$ are isomorphic and no two of the 3-manifolds $X(i)$ are even of the same homotopy type, and we see that there are infinitely many different affine equivalences (or even homotopy classes) of three-dimensional locally affine spaces.

The situation with respect to the analogue of the Second Theorem is unknown, but it is a part of a larger problem which we discuss below.

One possible way to try to patch the First Theorem would be to replace the conclusion that π have a free abelian subgroup of finite index with the conclusion that π has a solvable subgroup of finite index. (See [5], but beware that Lemma 2 is false. See also [4] (with care).) Another approach is taken in [36].

B. *Almost Flat Manifolds*

Another generalization of the Bieberbach Theorems is Gromov's theory of almost flat manifolds (see [38] and [14]).

Definition 6.2: An *almost flat manifold* is a manifold X s.t. $\forall\epsilon > 0, X$ has a riemannian structure with the property that for all $x \in X$ and for all $U_x, V_x \in X_x$ of length one and with $\langle U_x, V_x\rangle_x = 0$, we have

$$|\langle R(U_x, V_x) \cdot V_x, U_x\rangle_x| < \epsilon.$$

(The number $\langle R(U_x, V_x) \cdot V_x, U_x\rangle_x$ is called the *sectional curvature* of the plane spanned by U_x and V_x in X_x.)

Gromov has proved

Theorem 6.1 : *Let X be an almost flat manifold. Then $\pi_1(X)$ has a normal nilpotent subgroup of finite index.*

Gromov has said that he came across this theorem by trying to understand Bieberbach's First Theorem, so it was reasonable to expect his methods to lead to a new proof of the First Theorem. Peter Büser has recently published such a proof ([13]).

C. *The Topological Space Form Problem*

Bieberbach's Second Theorem (Theorem 5.4) tells that any two (compact) flat manifolds with the same fundamental group are somewhat better than diffeomorphic. A similar theorem is true for compact nilmanifolds which are differential manifolds that admit the transitive action of a nilpotent Lie group. Flat manifolds and nilmanifolds share the property of being covered by a contractible space. Such a space is said to be *aspherical.* A reasonable aspherical space X has no higher homotopy, i.e. $\pi_i(X) = 0$ for $i > 1$.

One might conjecture that compact asperical manifolds are classified up to diffeomorphism by their fundamental groups. One might, but one would be wrong. The trouble lies in the differential structure. You can change the differential structure of a manifold without changing the fact that it is compact and aspherical. One way to try and change the differential structure of a manifold is to take the connected sum with a "bad" sphere, i.e. a sphere with a non-standard differential structure. In general, this may or may not change the differential structure. However in the case of the n-torus T^n this always works. The differential structure always changes. So it is hopeless to expect two compact aspherical manifolds with the same fundamental groups to be diffeomorphic.

The "right" conjecture is the following:

Open Problem 6.1: Any two compact aspherical manifolds with isomorphic fundamental groups are homeomorphic.

"Right" means that no counterexample is currently known. In fact some parts of this open problem have been proven, and it is actively being worked on. This is the so-called topological space form problem.

The best results known at the present writing are due to Farrell and Hsiang (see [32]). Of interest to us here are the following theorems:

Theorem 6.2 : *Let X be an n-dimensional (compact) flat manifold with $n \neq 3,4$ and let Y be an aspherical manifold with $\pi_1(X) \approx \pi_1(Y)$. Then X and Y are homeomorphic.*

Theorem 6.3 : *Let X be a compact n-dimensional manifold with $n \neq 3,4$. Then if $\pi_1(X)$ has a nilpotent subgroup of finite index, X is almost flat.*

Their techniques are considerably more advanced than any used in this book and depend on deep results in the theory of topology of manifolds.

D. *The Holonomy Group of Compact Riemannian Manifolds*

The Holonomy Theorem (Theorem 3.2) tells us that the tangent space of the holonomy group of a flat manifold is trivial, so the holonomy group of a flat manifold is a totally disconnected subgroup of O_n (if we pick an orthonormal basis of X_n, $\Phi(X, x)$ can be considered as a subgroup of O_n). One of the remarkable consequences of the First Theorem is that if the manifold is compact, the holonomy group is finite. If we knew that the holonomy group of a compact riemannian manifold were a closed subgroup of O_n, then we could use Theorem 3.2 to conclude that it is a discrete subgroup of O_n and hence a finite subgroup of O_n since O_n is compact. Thus we could avoid the First Theorem. So we propose the following:

Open Problem 6.2: The holonomy group of a compact riemannian manifold is a closed subgroup of the orthogonal group.

It turns out, however, that properly stated, this open problem is itself the natural extension of the First Theorem to riemannian manifolds that are not necessarily flat. Before we "properly state" it, let us remark that Example 2.5 of Chapter I (the "infinite screw") shows compactness is essential. Now let X be a compact riemannian manifold. There is a theorem of Borel and Lichnerowicz (see [49], page186) which says that $\Phi_0(X)$ is a closed subgroup of O_n. The ideas in the proof of this theorem are as follows:

First, note that $\Phi_0(X) \approx \Phi_0(\tilde{X})$ where $\tilde{X}$ is the universal covering space of X.

Second, it is a standard fact that the holonomy group of a simply connected manifold breaks up into a product $\Phi^1 \times \Phi^2 \times \cdots \times \Phi^k$, and each Φ^i acts irreducibly on a euclidean space of the appropriate dimension.

Third and finally, there is a theorem which says that a subgroup of O_j which acts irreducibly on $\mathbb{R}^j$ is a closed subgroup. Hence the question of whether $\Phi(X)$ is closed in O_n is equivalent to the question of whether $\Phi(X)$ has a finite number of components, i.e. if $\Phi(X)/\Phi_0(X)$ is finite. So now the open problem is beginning to look a little more like the First Theorem. We can do even better.

Definition 6.3: Let $I(X)$ be the group of isometries mapping X to X.

There is a famous theorem of Myers and Steenrod which asserts that $I(X)$ is a Lie group. We know, for example, that $I(\mathbb{R}^n) = \mathcal{M}_n$.

Let $\tilde{X}$ be a simply connected riemannian manifold. Let $\tilde{x}_0 \in \tilde{X}$ and $\alpha \in I(\tilde{X})$. Let c be any curve from $\tilde{x}$ to $\alpha \cdot \tilde{x}_0$. Define a map $r_c : \tilde{X}_{\tilde{x}_0} \to \tilde{X}_{\tilde{x}_0}$ by

$$r_c(\alpha) = (\|_0^1)^{-1} \circ d\alpha_{\tilde{x}_0}.$$

In words, you use the differential of α to take a vector from $\tilde{x}_0$ to $\alpha\tilde{x}_0$, and then parallel translate it back to $\tilde{x}_0$ along c. Let $O(\tilde{X}_{\tilde{x}_0})$ be the orthogonal group of the tangent space. So $\Phi(\tilde{X}, \tilde{x}_0)$ (which is $\Phi_0(\tilde{X}, \tilde{x}_0)$) is a subgroup of $O(\tilde{X}_{\tilde{x}_0})$.

Exercise 6.2: Show that $r_c(\alpha)$ is independent of c mod $\Phi(\tilde{X}, \tilde{x}_0)$.

Hence we have defined a map

$$r : I(\tilde{X}) \to O(\tilde{X}_{\tilde{x}_0})/\Phi(\tilde{X}, \tilde{x}_0).$$

However, $O(\tilde{X}_{\tilde{x}_0})/\Phi(\tilde{X}, \tilde{x}_0)$ is not necessarily a group. In analogy with the map $r : M_n \to O_n$ in the flat case, we want a map into a group. And we can get one.

Exercise 6.3: Let N be the normalizer of $\Phi(\tilde{X}, \tilde{x}_0)$ in $O(\tilde{X}_{\tilde{x}_0})$. Show that

$$r_c(\alpha) \in N \quad \forall \alpha \in I(X).$$

Now $N/\Phi(\tilde{X}, \tilde{x}_0)$ *is* a group.

Definition 6.4: Let $\tilde{X}$ be a simply connected riemannian manifold. The *orthogonal group of* $\tilde{X}$ is defined by

$$O(\tilde{X}) = O(\tilde{X}, \tilde{x}_0) = N/\Phi(\tilde{X}, \tilde{x}_0).$$

The *rotation map* of $\tilde{X}$ is the map $r : I(X) \to O(\tilde{X})$ defined above.

It is an easy exercise to see that if $\tilde{X}$ is $\mathbb{R}^n$, then $I(X)$ is M_n, and $O(\tilde{X})$ is O_n, and r is the usual rotation map.

Let X be any riemannian manifold, and let $\tilde{X}$ be its simply connected covering space. Then $\pi_1(X)$ acts as a group of isometries on $\tilde{X}$. More precisely, if we give $\tilde{X}$ the riemannian structure which makes the covering map a local isometry, then the fundamental group acts isometrically on $\tilde{X}$. Hence $\pi_1(X)$ can be considered as a subgroup of $I(\tilde{X})$.

Exercise 6.4: Show that $r(\pi_1(X))$ is isomorphic to $\Phi(X)/\Phi_0(X)$.

Therefore the open problem of this section can be restated as the following:

Open Problem 6.3: $r(\pi_1(X))$ is finite if X is compact.

This formulation leads to many interesting questions which arise by analogy with the flat case. Is r surjective? Let $K = \ker(r \mid \pi_1(X))$. What can be said about the elements in K? Are they "like" pure translations? What is the algebraic structure of K? Is $\tilde{X}/K$ compact?

There is an entirely different approach to the open problem which proceeds by looking at examples. First of all, a paper of Wolf ([78]) shows the open problem is true for symmetric spaces. For non-symmetric spaces, Berger's thesis ([8]) gives a list of all the possible irreducible identity components for holonomy groups of riemannian manifolds. By irreducible we mean, of course, not reducible. A reducible subgroup Φ of O_n is one for which you can find coordinates in which all the matrices in Φ take the form

$$\begin{pmatrix} * & 0 \\ 0 & ** \end{pmatrix}.$$

What we can do is to look at Berger's list and reason as follows: If Φ_0 is on the list, then any holonomy group Φ with Φ_0 as identity component must be contained in the normalizer of Φ_0 in O_n. Call this normalizer N. Hence Φ/Φ_0 must be a subgroup of N/Φ_0. If N/Φ_0 turns out to be finite, then no riemannian manifold whose holonomy group has Φ_0 as identity component can be a counterexample to our open problem. Berger actually computes N/Φ_0, and it *is* finite except for the last two groups on the list, SU_n and Sp_n (see chapter I of [24] for a description of these groups). For these groups, we have

$$N/SU_n \approx \mathbb{Z}_2 \times S^1 \text{ and } N/Sp_n \approx S^3,$$

where S^1 is the circle group (i.e. the group of complex numbers of norm one) and S^3 is the group of unit quaternions.

Using methods due to Cheeger and Gromoll ([23]), these cases can be handled, i.e. no manifold whose holonomy group has identity component SU_n or Sp_n can be a counterexample to our open problem. This finishes the irreducible case.

Actually, many reducible cases can be handled also. The best result at the present writing is the following:

Theorem 6.4 (Charlap, Cheeger, and Gromoll): *Let X be a compact riemannian manifold whose holonomy group is not closed in O_n. Then the universal covering space $\tilde{X}$ of X breaks into a product*

$$\tilde{X} = A \times B \times C,$$

where A has holonomy group SU_n or Sp_n and B is $\mathbb{R}^k$ for $k \geq 2$ (and C is what's left over).

One final word on this problem is that no concrete examples of compact riemannian manifolds with SU_n or Sp_n are known. A famous conjecture of Calabi recently proved by Yau ([80]) implies that any Kähler manifold with vanishing first Chern class has a riemannian structure with SU_n as holonomy group. But this is an existence theorem, and although there are many such manifolds known, nobody has yet been able to construct the right riemannian structure on any of them.

More recently, Bogomolov has claimed to have proved there were *no* riemannian manifolds with Sp_n as holonomy group, but there was an error in his proof. For information concerning this and related matters, see [7].

Chapter III

Classification Theorems

1. The Algebraic Structure of Bieberbach Groups

We have defined Bieberbach subgroups of $\mathcal{M}_n$ as the torsionfree, discrete, uniform ones. We have seen that such a subgroup π contains a free abelian subgroup $\pi \cap \mathbb{R}^n$ which is normal, maximal abelian, and of finite index in π. We now define what it means for an abstract group to be Bieberbach.

Definition 1.1: A *crystallographic* group is a group which contains a finitely generated maximal abelian torsionfree subgroup of finite index. A *Bieberbach group* is a crystallographic group which is itself torsionfree.

Exercise 1.1: i) Show that the maximal abelian subgroup of a crystallographic group is normal. (This observation was pointed out by F. T. Farrell.)

ii) Show that any finitely generated torsionfree group with a free abelian subgroup of finite index is a Bieberbach group. (**Hint:** Look at the subgroup of element that have finitely many conjugates.)

The rank of this unique subgroup is called the *dimension* of the crystallographic group.

Some Remarks on Terminology: We first used the term "Bieberbach group" in [17], but there we used it to mean what we call here "crystallographic subgroup of $\mathcal{M}_n$." In [17], "crystallographic group" was defined to be a discrete uniform subgroup of $\mathcal{M}_3$. What we call "Bieberbach subgroup of $\mathcal{M}_n$" here, was called "torsionfree Bieberbach group" in [17]. Finally what we have just defined as "Bieberbach group" in the above definition was called "abstract torsionfree Bieberbach group" in [17].

What can be the reasons behind this dreadful confusion? Firstly, there were good reasons for the terminology in [17]. As we have seen, the groups studied by Bieberbach were the discrete uniform subgroups of

$\mathcal{M}_n$. The groups studied by the crystallographers were the discrete uniform subgroups of $\mathcal{M}_3$. The hypothesis that the group be torsionfree only appears when the study of flat manifolds was initiated, much later than Bieberbach, let alone the crystallographers. However, the term "torsionfree Bieberbach group" is admittedly clumsy, and subsequent authors, particularly Farrell and Hsiang, began using the terminology we use here, i.e. "crystallographic" is discrete and uniform, and "Bieberbach" is torsionfree crystallographic. So what we are doing here is giving in to the currently accepted usage in spite of its historical inaccuracy.

The distinction between "abstract Bieberbach groups" and "Bieberbach subgroups of $\mathcal{M}_n$" is invidious, as was remarked in [17], and will be demonstrated here. Also we used to require in the definition of Bieberbach group that the abelian subgroup be normal, but as remarked in the above exercise, F. T. Farrell pointed out that this could be proved from the rest of the definition. This observation also appears to have been known to E. Calabi.

We already know that a Bieberbach subgroup of $\mathcal{M}_n$ is a Bieberbach group, so we must show that a Bieberbach group is a Bieberbach subgroup of $\mathcal{M}_n$. Actually, it is a theorem of L. Auslander and M. Kuranishi ([6]) that a crystallographic group is a crystallographic subgroup of $\mathcal{M}_n$, and that certainly suffices.

Theorem 1.1 (Auslander and Kuranishi): *Let π be a crystallographic group of dimension n . Then there is a monomorphism $F : \pi \to \mathcal{M}_n$ s.t. $F(\pi)$ is a crystallographic subgroup of $\mathcal{M}_n$.*

Proof: Let M be the unique normal maximal abelian subgroup of π defined in Exercise 1.1. Since π has dimension n, M is free of rank n. Let $b_1, \ldots, b_n$ be a basis for M, and define $\tilde{F} : M \to \mathbb{R}^n$ by $\tilde{F}(b_i) = e_i$ for $i = 1, 2, \ldots, n$, where $e_1, \ldots, e_n$ is the usual basis of $\mathbb{R}^n$. Let $\sigma \in \Phi = \pi/M$. Φ acts on M, so we can write

$$\sigma \cdot b_i = \sum_j \sigma_{ij} b_j, \text{ with } \sigma_{ij} \in \mathbb{Z},$$

and the matrix (σ_{ij}) will be in GL_n. We can then define $F' : \Phi \to GL_n$ by $F'(\sigma) = (\sigma_{ij})$. Now let $f : \Phi \times \Phi \to M$ be a 2-cocycle corresponding to the extension

$$0 \longrightarrow M \longrightarrow \pi \longrightarrow \Phi \longrightarrow 1, \tag{1}$$

and consider π to be $\Phi \times N$ with the multiplication

$$(\sigma, m)(\tau, n) = (\sigma\tau, \sigma \cdot n + m + f(\sigma, \tau)).$$

Now $F \circ f : \Phi \times \Phi \to \mathbb{R}^n$ is a 2-cocycle in $\mathbb{Z}^2(\Phi; \mathbb{R}^n)$, but since $\mathbb{R}^n$ is divisible, by Corollary 5.2, part ii) of Chapter I, $H^2(\Phi; \mathbb{R}^n) = 0$, so there exists a 1-cochain $g \in C^1(\Phi; \mathbb{R}^n)$ s.t. $\delta g = F \circ f$, i.e.

$$\delta g(\sigma, \tau) = \sigma \cdot g(\tau) - g(\sigma\tau) + g(\sigma) = \tilde{F} \circ f(\sigma, \tau). \tag{2}$$

Define $F : \pi \to A_n$ by

$$F(\sigma, m) = \left(F'(\sigma), \tilde{F}(m) + g(\sigma)\right). \tag{3}$$

Although F as defined doesn't take values in $\mathcal{M}_n$, since $F'(\Phi)$ is finite, by Proposition 6.2, we can, by a change of basis, assume $F'(\Phi)$ is in O_n, not merely GL_n. Now we can assume $F : \pi \to \mathcal{M}_n$.

Claim: F is a homomorphism.

For this computation, we simplify the notation by identifying $F'(\sigma)$ with σ and $\tilde{F}(m)$ with m, so (3) becomes

$$F(\sigma, m) = (\sigma, m + g(\sigma)). \tag{4}$$

Now if $(\tau, n) \in \pi$, then $(\tau, n)^{-1} = (\tau^{-1}, -\tau^{-1} \cdot n - f(\tau^{-1}, \tau))$, so

$$\begin{aligned} F((\sigma, m) \cdot (\tau, n)^{-1}) &= F\left(\sigma\tau^{-1}, -\sigma\tau^{-1} \cdot n - \sigma \cdot f(\tau^{-1}, \tau) + m + f(\sigma, \tau^{-1})\right) \\ &= \left(\sigma\tau^{-1}, -\sigma\tau^{-1} \cdot n - \sigma \cdot f(\tau^{-1}, \tau) + m + f(\sigma, \tau^{-1}) + g(\sigma\tau^{-1})\right) \end{aligned}$$

by (4). While

$$\begin{aligned} [F(\sigma, m)] \cdot [F(\tau, n)]^{-1} &= (\sigma, m + g(\sigma)) \cdot (\tau, n + g(\tau))^{-1} \\ &= \left(\sigma\tau^{-1}, -\sigma\tau^{-1} \cdot n - \sigma\tau^{-1} \cdot g(\tau) + m + g(\sigma)\right) \end{aligned}.$$

To prove the claim, it suffices to show

$$-\sigma \cdot f(\tau^{-1}, \tau) + f(\sigma, \tau^{-1}) + g(\sigma\tau^{-1}) = -\sigma\tau^{-1} \cdot g(\tau) + g(\sigma).$$

By (2), $f(\sigma, \tau^{-1}) = \sigma \cdot g(\tau^{-1}) - g(\sigma\tau) + g(\sigma)$, so now we must show

$$-\sigma \cdot f(\tau^{-1}, \tau) = -\sigma\tau^{-1} \cdot g(\tau) - \sigma \cdot g(\sigma)$$

or

$$f(\tau^{-1},\tau) = \tau^{-1} \cdot g(\tau) + g(\sigma).$$

But $\delta g(\tau^{-1},\tau) = \tau^{-1} \cdot g(\tau) - g(1) + g(\tau^{-1}) = f(\tau^{-1},\tau)$, since we are using normalized cochains, so $g(1) = 0$, and the claim follows.

Since M is maximally abelian, F' is injective. Thus if $F(\sigma, m) = (I, 0)$ then $\sigma = 1$. Since $g(1) = 0$, (3) implies that $G(m) = 0$, and since G is clearly injective, $m = 0$. Therefore F is a monomorphism.

Now $F(M) = \mathbb{Z}^n \subset \mathbb{R}^n$ (or more precisely, it is $\mathbb{Z}^n$ after a change of basis), so $F(M)$ is discrete, and since $F(\pi)/F(M)$ is isomorphic to Φ which is finite, $F(\pi)$ is discrete in $\mathcal{M}_n$.

To see that $\mathbb{R}^n/F(\pi)$ is compact, note that it is the continuous image of $\mathbb{R}^n/F(M)$ which is a torus which is compact. (Of course, we know that $\mathbb{R}^n/F(\pi)$ is actually covered by the torus T^n.) Hence $F(\pi)$ is a crystallographic subgroup of $\mathcal{M}_n$. □

2. A General Classification Scheme for Bieberbach Groups

We know that a Bieberbach group π satisfies an exact sequence

$$0 \longrightarrow M \longrightarrow \pi \longrightarrow \Phi \longrightarrow 1 \tag{5}$$

with M free abelian (say, of rank n), Φ finite, and π torsionfree. By examining our proof of Bieberbach's Third Theorem, we can easily devise the following scheme for constructing and classifying such groups:

A) Pick a finite group Φ which can be the holonomy group of a Bieberbach group.

B) Take a faithful representation of Φ on a free abelian group M.

C) Find a coholomogy class $\alpha \in H^2(\Phi; M)$ s.t. the associated extension π is torsionfree.

D) See if this π is isomorphic to any other π' corresponding to any $\alpha' \in H^2(\Phi : M')$ where M' is any other Φ-module.

We are going to examine C) and D) in this section and prove some results which will be sufficient in some easy cases. In Section 4 of this chapter, we will show that any finite group Φ can be the holonomy group of a Bieberbach group (or a flat manifold). This will more or less dispose of A). We say "more or less" since we still don't know, given Φ, the possible dimensions in which it appears as a holonomy group. B) is by far the most difficult step, and is equivalent to a whole branch of mathematics, "integral representations of finite groups." As we remarked previously, we usually sidestep B) by only looking at those groups Φ whose integral representations are known,

First we look at C). We must examine what happens to cohomology when we restrict our attention to subgroups of Φ. Let G and H be groups, M an H-module, and $F : G \to H$, a homomorphism. Then, in the usual way, we can make M into a G-module by setting $g \cdot m = F(g) \cdot m$ for $g \in G$, and $m \in M$. We denote this G-module by $F^{-1}(M)$. We can use this construction to define a homomorphism $F^{\#} : C^2(H; M) \to C^2(G; F^{-1}(M))$ by

$$\left[F^{\#}(c)\right](g_1, g_2) = c(F(g_1), F(g_2))$$

for $c \in C^2(H; M)$ and $g_1, g_2 \in G$. If N is another H-module and $f : M \to N$ is a homomorphism of H-modules, then we can define another homomorphism $f_{\#} : C^2(H; M) \to C^2(H; N)$ by

$$[f_{\#}(c)](h_1, h_2) = f(c(h_1, h_2))$$

for $c \in C^2(H; M)$ and $h_1, h_2 \in H$.

Exercise 2.1: Show that if $c \in Z^2(H; M)$ (respectively $B^2(H; M)$), then $F^{\#}(c) \in Z^2(G; F^{-1}(M))$ (respectively $B^2(G; F^{-1}(M))$), and $f_{\#}(c) \in Z^2(H; N)$ (respectively $B^2(H; N)$).

Therefore, $F : G \to H$ induces a homomorphism

$$F^* : H^2(H; M) \to H^2(G; F^{-1}(M)),$$

and $f : M \to N$ induces a homomorphism

$$f_* : H^2(H; M) \to H^2(H; N).$$

If you know about category theory, you see this shows that for a fixed group G, $M \mapsto H^2(G; M)$ is a covariant functor from the category of G-modules to the category of abelian groups. H^2 is *not*, in any natural way, a functor on the first variable, i.e. on the category of groups.

Definition 2.1: We call $F^{-1}(M)$ the G-module *induced* by F. If $I : G \hookrightarrow H$ is an inclusion (or monomorphism), we call $I^* : H^2(H; M) \to$

$H^2(G; I^{-1}(M))$ *restriction.* In this case, we usually identify $I^{-1}(M)$ with M (they are certainly hard to tell apart).

Theorem 2.1 : *Let π be the extension corresponding to $\alpha \in H^2(\Phi; M)$. Then π is torsionfree iff for each injection $I : \mathbb{Z}_p \hookrightarrow \Phi$ of a group of prime order into Φ, $I^*(\alpha) \neq 0$.*

Proof: Suppose $I^*(\alpha) = 0$. Then we can choose $f \in \alpha$ s.t. $f(\sigma, \tau) = 0$ whenever $\sigma, \tau \in I(\mathbb{Z}_p) \subset \Phi$. This means we can find a section $s : \Phi \to \pi$ s.t.

$$s(\sigma, \tau) = s(\sigma) \cdot s(\tau) \quad \forall \sigma, \tau \in I(\mathbb{Z}_p).$$

Hence $s \circ I(\mathbb{Z}_p)$ is a subgroup of π of order p, so π is not torsionfree.

Conversely, suppose π has an element of finite order. Then π has an element of prime order, say $(\sigma, m)^p = (1, 0)$ (we are, as usual, thinking of π as the set $\Phi \times M$).

Exercise 2.2: Show

$$(\sigma, m)^p = \left(\sigma^p, \left(\sum_{k=0}^{p-1} \sigma^k \right) \cdot m + \sum_{k=1}^{p-1} f(\sigma^k, \sigma) \right).$$

So we see that the subgroup ρ of Φ generated by σ is cyclic of order p. We want to show $I^*([f]) = 0$, where $[f]$ is the cohomology class of the extension (5) and $f \in [f]$, and $I : \rho \hookrightarrow \Phi$ is the inclusion. Since $(\sigma, m)^p = (1, 0)$, the above exercise shows that

$$\sum_{k=1}^{p-1} f(\sigma^k, \sigma) \in \Sigma \cdot M \tag{6}$$

where Σ is the element of $\mathbb{Z}[\rho](\subset \mathbb{Z}[\pi])$ defined by

$$\Sigma = \sum_{k=0}^{p-1} \sigma^k = \sum_{\sigma \in \rho} \sigma.$$

If we confuse $I^{\#}(f)$ with f (and we will), we want to find $g \in C^1(\rho; M)$ s.t. $\delta g = f$, i.e.

$$f(\sigma^k, \sigma^j) = \sigma^k \cdot g(\sigma^j) - g(\sigma^{k+j}) + g(\sigma^k). \tag{7}$$

Define g by

$$g(\sigma^k) = -\left[\sum_{i=1}^{k-1} f(\sigma^i,\sigma)\right] + \left(\sum_{i=0}^{k-1}\sigma^i\right)\cdot m,$$

where m is defined via (6) by

$$\sum_{k=1}^{p-1} f(\sigma^k,\sigma) = \Sigma\cdot m.$$

Then

$$\begin{aligned}
\delta g(\sigma^k,\sigma) &= \sigma^k\cdot g(\sigma) - g(\sigma^{k+1}) + g(\sigma^k)\\
&= \sigma^k\cdot m + \sum_{i=1}^{k} f(\sigma^i,\sigma) - \left(\sum_{i=0}^{k}\sigma^i\right)\cdot m\\
&\quad - \sum_{i=1}^{k-1} f(\sigma^i,\sigma) + \left(\sum_{i=0}^{k-1}\sigma^i\right)\cdot m\\
&= \sigma^k\cdot m + f(\sigma^k,\sigma) - \sigma^k\cdot m\\
&= f(\sigma^k,\sigma).
\end{aligned}$$

Therefore (7) holds for $j = 1$. Now assume inductively (on j) that $\delta g(\sigma^k,\sigma^i) = f(\sigma^k,\sigma^i)$ for all k and for all $i \le j$. Since $\delta\delta g = 0$,

$$0 = \delta\delta g(\sigma^k,\sigma^j,\sigma) = \sigma^k\cdot\delta g(\sigma^j,\sigma) - \delta g(\sigma^{k+j},\sigma) + \delta g(\sigma^k,\sigma^{j+1}) - \delta g(\sigma^k,\sigma^j),$$

and we get

$$\begin{aligned}
\delta g(\sigma^k,\sigma^{j+1}) &= \delta g(\sigma^k,\sigma^j) - \sigma^k\cdot\delta g(\sigma^j,\sigma) + \delta g(\sigma^{k+j},\sigma)\\
&= f(\sigma^k,\sigma^j) - \sigma^k\cdot f(\sigma^j,\sigma) + f(\sigma^{k+j},\sigma) \quad ,\\
&= f(\sigma^k,\sigma^{j+1})
\end{aligned}$$

where the second line follows by induction, and the third line follows by writing out $\delta f(\sigma^k,\sigma^j,\sigma) = 0$. Hence $\delta g = f$ and $I^*(\alpha) = 0$. □

Our next theorem will be very helpful in looking at problem D), namely determining whether two extensions are isomorphic. The earliest reference we know for this theorem is [I], although the result is simple enough to have been considered "well-known" before 1965. It has since been rediscovered in a number of papers including some unlikely contexts (see [A&J]).

Definition 2.2: Let M and N be Φ-modules. A *semi-linear homomorphism* from M to N is a pair (f, A) with $f \in \mathrm{Hom}_{\mathbb{Z}}(M, N)$ (i.e. $f(m+n) = f(m) + f(n)$), and $A \in \mathrm{Aut}(\Phi)$ s.t. $f(\sigma \cdot m) = A(\sigma) \cdot f(m)$ for $\sigma \in \Phi$ and $m \in M$. We let $\mathrm{Hom}_S(M, N)$ be the set of all semi-linear homomorphisms from M to N.

Let $(f, A) \in \mathrm{Hom}_S(M, N)$ and $\alpha \in H^2(\Phi; M)$. Then if $c \in \alpha$ (so $c \in Z^2(\Phi; M)$), we can still define $f_\#(c) : \Phi \times \Phi \to N$ by

$$[f_\#(c)](\sigma_1, \sigma_2) = f[c(\sigma_1, \sigma_2)].$$

Exercise 2.3: Show $f_\#(c) \in Z^2(\Phi; A^{-1}(N))$, so (f, A) defines a homomorphism from $H^2(\Phi; M)$ to $H^2(\Phi; A^{-1}(N))$ which by an "abuse of terminology," we again denote by f_*.

Theorem 2.2 : *Let M and N be faithful Φ-modules, $\alpha \in H^2(\Phi; M)$, and $\beta \in H^2(\Phi; N)$. Let π (respectively ρ) be the extension corresponding to α (respectively β). Then π is isomorphic to ρ iff there is $(f, A) \in \mathrm{Hom}_S(M, N)$ such that f is bijective and*

$$f_*(\alpha) = A^*(\beta). \tag{8}$$

Proof: Suppose we are given $(f, A) \in \mathrm{Hom}_S(M, N)$ with f bijective s.t. (8) holds. Define a map $F : \pi \to \rho$ by

$$F(\sigma, m) = (A(\sigma), f(m))$$

where, as usual, we think of π and ρ as $\Phi \times M$ (as sets). Now pick cocycles $a \in \alpha$ and $b \in \beta$ s.t.

$$f[a(\sigma, \tau)] = b(A(\sigma), A(\tau)).$$

We can do this because of (8). F is clearly bijective, and

$$\begin{aligned} F[(\sigma, m) \cdot (\tau, n)] &= F(\sigma\tau, \sigma \cdot n + m + a(\sigma, \tau)) \\ &= (A(\sigma\tau), f(\sigma \cdot n) + f(m) + f(a(\sigma, \tau))) \\ &= (A(\sigma)A(\tau), A(\sigma) \cdot f(n) + f(m) + b(A(\sigma), A(\tau))) \\ &= (A(\sigma), f(m)) \cdot (A(\tau), f(n)) \\ &= F(\sigma, m) \cdot F(\tau, n), \end{aligned}$$

so F is an isomorphism (of groups).

Now suppose $F : \pi \to \rho$ is an isomorphism. Then by Exercise 1.1, $F(M) = N$, so F induces an isomorphism $A : \pi/M \to \pi/N$, but both π/M and π/N are Φ, so $A \in \mathrm{Aut}(\Phi)$. Now let $f = F \mid M$. We must show

i) $f(\sigma \cdot m) = A(\sigma) \cdot f(m)$, and

ii) $f_*(\alpha) = A^*(\beta)$.

Let $p : \pi \to \Phi$ and $p' : \rho \to \Phi$ be the projections. Let $x \in \pi$ and suppose $p(x) = \sigma \in \Phi$. Then by the definition of A, $p'(F(x)) = A(\sigma)$, so

$$f(\sigma \cdot m) = F(xmx^{-1}) = F(x)F(m)F(x)^{-1} = A(\sigma) \cdot f(m).$$

Now let $s : \Phi \to \pi$ be a section of π. Then $s' : \Phi \to \rho$ defined by $s'(\sigma) = F \circ s \circ A^{-1}(\sigma)$ is a section of ρ. Let α and β be represented by the 2-cocycles a and b, respectively, which are defined by

$$a(\sigma, \tau) = s(\sigma)s(\tau)s(\sigma\tau)^{-1},$$

and

$$b(\sigma, \tau) = s'(\sigma)s'(\tau)s'(\sigma\tau)^{-1}.$$

Then

$$\begin{aligned}
\left[A^{\#}(b)\right](\sigma, \tau) &= b(A(\sigma), A(\tau)) \\
&= s'(A(\sigma))s'(A(\tau))s'(A(\sigma\tau))^{-1} \\
&= F(s(\sigma)) \cdot F(s(\tau)) \cdot F(s(\sigma\tau)^{-1}) \\
&= F(a(\sigma, \tau)) \\
&= [F_{\#}(a)](\sigma, \tau),
\end{aligned}$$

which yields ii). □

3. Digression — Cohomology of Groups

We need to know some more about the cohomology of groups before we can present the solution to problem A), which is another theorem of L. Auslander and M. Kuranishi. This theorem states that any finite group can be the holonomy group of a Bieberbach group (or flat manifold). Although, again, we could get along with presenting just what we need in an ad hoc fashion, at some point in this book we will need the general theory of the cohomology of groups, so we may as well do it here.

Fix a (not necessarily finite) group Φ. When we talk about modules in this section, we will mean Φ-modules. Although the notion of "free module" could suffice for our purposes, there is a generalization which turns out to be more natural for the general theory.

Definition 3.1: A module L is said to be *projective* if whenever there is a homomorphism $f : L \to N$ and an epimorphism $p : M \to N$, there is a homomophism $\bar{f} : L \to M$ s.t. $p \circ \bar{f} = f$. In terms of diagrams, there always exists a homomorphism along the diagonal arrow making the following diagram commutative:

$$\begin{array}{ccccc} & & L & & \\ & \swarrow & \downarrow f & & \\ M & \xrightarrow{p} & N & \longrightarrow & 0. \end{array}$$

Roughly speaking, L is projective iff every map to a quotient module can be lifted to the overlying module.

Proposition 3.1: *L is projective iff it is a direct summand of a free module.*

Proof: It is trivial to see that a free module is projective and that a direct summand of a projective module is projective. Hence a direct summand of a free module is projective.

On the other hand, let L be projective. Write $L = F/R$ where F is a free module and R is an appropriate submodule of F. Then we have $p : F \to L$ surjective, and by considering the identity map of L,

$$\begin{array}{ccccc} & & L & & \\ & \swarrow & \downarrow & & \\ F & \xrightarrow{p} & L & \longrightarrow & 0. \end{array}$$

we get a map $f : L \to F$ s.t. $p \circ f$ is the identity, which implies that L is a direct summand of F. □

Example 3.1: Let R be the ring $\mathbb{Z}_6$, and let M (respectively N) be the R-module $\mathbb{Z}_2$ (respectively $\mathbb{Z}_3$).

Exercise 3.1: i) Show $R = M \oplus N$.

ii) Show M is projective.

iii) Show M cannot be free. (**Hint:** Count.)

iv) This example is not an example of a module over a group (or group ring, to be precise). Can you find an example of a group Φ and a Φ-module M which is projective but not free. (**Hint:** Look at the remark following Theorem 3.1 of Chapter IV.)

Definition 3.2: Let M be a module. A *complex* X over M is a sequence of modules and homomorphisms

$$\dots \longrightarrow X_n \xrightarrow{d_n} X_{n-1} \xrightarrow{d_{n-1}} X_{n-2} \longrightarrow \dots \longrightarrow X_1 \xrightarrow{d_1} X_0 \xrightarrow{\epsilon} M \tag{9}$$

with the properties that

i) ϵ is surjective,

ii) $d_1 \circ \epsilon = 0$, and

iii) $d_n \circ d_{n-1} = 0$ for $n = 1, 2, \dots$.

The map d_n is called the nth *differential* of X, and ϵ is called the *augmentation* of X. A complex over M is said to be a *resolution of M for Φ* if the sequence (9) is exact. If each X_n is free (respectively projective), we say X is *free* (respectively *projective*). Any free complex is, of course, projective.

Proposition 3.2: *For any Φ and M, there is a free resolution of M for Φ.*

Proof: Let X_0 be any free module that maps onto M, and let $\epsilon : X_0 \to M$ be that epimorphism. Let $K_0 = \ker \epsilon$, let X_1 be any free module that maps onto K_0, and let $p : X_1 \to K_0$ be that epimorphism. Let $i : K_0 \hookrightarrow X_0$ be the inclusion, and put $d_1 = i \circ p$.

$$\begin{array}{ccccccc} X_1 & \xrightarrow{d_1} & X_0 & \xrightarrow{\epsilon} & M & \longrightarrow & 0 \\ & \searrow & & \nearrow & & & \\ & & K_0 & & & & \\ & \nearrow & & \searrow & & & \\ 0 & & & & 0. & & \end{array}$$

Then the top row of the above diagram is exact, and we clearly can continue in this fashion (use induction, if you insist). □

So now we have defined these rather complicated (or complex) objects called complexes and have shown they always exist. What in the world are they good for? They provide the natural setting to study cohomology (and homology). Before we get into that though, we want to be able to compare complexes, i.e. we want to define maps between them (actually to make them into a category).

Definition 3.3: Let X and Y be complexes with differentials d_n and ∂_n, respectively. A *chain map* $f : X \to Y$ is a sequence $\{f_i\}$ with $f_i \in \mathrm{Hom}(X_i, Y_i)$ for $i = 0, 1, 2, \dots$ s.t.

$$\partial_i f_i = f_{i-1} d_i \text{ for } i = 1, 2, \dots$$

The group of all chain maps from X to Y will be denoted by $\mathrm{Hom}(X,Y)$.

If $f, g \in \mathrm{Hom}(X,Y)$, we say f and g are *homotopic* or *chain homotopic* if there is a sequence $\{s_i\}$ with $s_i \in \mathrm{Hom}(X_i, Y_{i+1})$ for $i = 0, 1, 2, \ldots$ s.t.

$$\partial_{i+1} s_i + s_{i-1} d_i = f_i - g_i \text{ for } i = 1, 2, \ldots$$

and

$$\partial_1 s_0 = f_0 - g_0.$$

Note that as we have defined things, chain maps and homotopies have nothing to do with the augmentation ϵ.

Definition 3.4: If X is a complex, we can define the *nth homology* of X to be the module

$$H_n(X) = \ker(d_n)/\mathrm{img}(d_{n+1}) \quad \text{for } n > 0,$$

and

$$H_0(X) = X_0/\mathrm{img}(d_1).$$

(Note that $\mathrm{img}(d_{n+1}) \subset \ker(d_n)$ since $d_{n+1} \circ d_n = 0$.) If $f \in \mathrm{Hom}(X,Y)$, we get a map $f_* : H_n(X) \to H_n(Y)$ by letting

$$f_*(\alpha) = [f(x)] \text{ for any } x \in \alpha \in H_n(X).$$

(Recall $[y]$ is the class of y in $H_n(Y)$.)

Exercise 3.2: i) Show that f_* is well-defined, i.e. if $x' \in \alpha$, then $[f(x')] = [f(x)]$.

ii) Show that if $f, g \in \mathrm{Hom}(X,Y)$ are homotopic, then $f_* = g_*$.

What we want to do is to show these complexes can be used to compute the cohomology of groups. The idea is that if our complexes are "reasonable," we can use any one for the computation, so we will pick one that is "good" for the group in question. The main required result is the following:

Theorem 3.1 : *Let M and N be modules and $F \in \mathrm{Hom}(M,N)$. Let X be a projective complex over M, and let Y be a resolution of N (for Φ). Then there exists $f \in \mathrm{Hom}(X,Y)$ s.t.*

$$\epsilon_Y \circ f_0 = F \circ \epsilon_X, \tag{10}$$

and any two such chain maps are homotopic.

We say f is a *lift* of F. The picture is the following:

$$\begin{array}{ccccccccccc}
\cdots & \longrightarrow & X_2 & \xrightarrow{d_2} & X_1 & \xrightarrow{d_1} & X_0 & \xrightarrow{\epsilon_X} & M & \longrightarrow & 0 \\
 & & \downarrow{\scriptstyle f_2} & & \downarrow{\scriptstyle f_1} & & \downarrow{\scriptstyle f_0} & & \downarrow{\scriptstyle F} & & \\
\cdots & \longrightarrow & Y_2 & \xrightarrow{\partial_2} & Y_1 & \xrightarrow{\partial_1} & Y_0 & \xrightarrow{\epsilon_Y} & N & \longrightarrow & 0.
\end{array}$$

Proof: Since X_0 is projective and ϵ_y surjective, $\exists f_0 : X_0 \to Y_0$ s.t. (10) holds. By induction, it suffices to construct $f_n : X_n \to Y_n$ given $f_{n-1}, f_{n-2}, \ldots, f_1, f_0$ which make the above diagram commute. Now

$$\partial_{n-1} \circ f_{n-1} \circ d_n = f_{n-2} \circ d_{n-1} \circ d_n = 0.$$

Therefore, $\mathrm{img}(f_{n-1} \circ d_n) \subset \ker(\partial_{n-1})$ which is $\mathrm{img}(\partial_n)$ since Y is exact (i.e. a resolution). So we have

$$\begin{array}{ccccc}
 & & X_n & & \\
 & & \downarrow{\scriptstyle f_{n-1}\circ d_n} & & \\
Y_n & \xrightarrow{\partial_n} & \mathrm{img}(\partial_n) & \longrightarrow & 0
\end{array}$$

and since X is projective, there is $f_n : X_n \to Y_n$ s.t. $\partial_n \circ f_n = f_{n-1} \circ d_n$ as desired.

Now given two lifts f and f' of F, we can construct a homotopy in pretty much the same fashion. However, the construction appears enough times in homological algebra (and in this book) so that we prefer to deduce it from the following:

Lemma 3.1: *Let X be a projective complex over M and Y a resolution of N. Suppose $f : X \to Y$ lifts $F : M \to N$. Let $T : M \to Y_0$ satisfy*

$$\epsilon_y \circ T = F. \tag{11}$$

Then $\exists s_i \in \mathrm{Hom}(X_i, Y_i)$ for $i = 0, 1, 2, \ldots$ s.t.

$$\partial_1 \circ s_0 + T \circ \epsilon_X = f_0, \tag{12}$$

and

$$\partial_{i+2} \circ s_{i+1} + s_i \circ d_{i+1} = f_{i+1} \text{ for } i = 0, 1, 2, \ldots \tag{13}$$

Proof: First we have

$$\epsilon_Y(f_0 - T\epsilon_X) = \epsilon_Y f - \epsilon_Y T\epsilon_X = F\epsilon_Y - F\epsilon_X = 0,$$

so $\operatorname{img}(f_0 - T\epsilon_X) \subset \ker\epsilon_Y = \operatorname{img}\partial_1$, i.e. we have the following commutative diagram:

$$\begin{array}{ccccc} & & X_0 & & \\ & & \big\downarrow{\scriptstyle f_0 - T\epsilon_X} & & \\ Y_1 & \xrightarrow{\partial_1} & \operatorname{img}(\partial_1) & \longrightarrow & 0. \end{array}$$

Since X_0 is projective, $\exists s_0 : X_0 \to Y_1$ s.t. $\partial_1 s_0 = f_0 - T\epsilon_X$, i.e. (12) holds. Let $T = s_1$, and suppose we have found $s_0, s_1, s_2, \ldots$ satisfying (13). We want $s_{i+1} : X_{i+1} \to Y_{i+2}$ with

$$\partial_{i+2} s_{i+1} = f_{i+1} - s_i d_{1+1}.$$

Now $\partial_{i+1}(f_{i+1} - s_i d_{i+1}) = f_{i+1} d_i - (f_i - s_{i-1} d_i) d_{i+1} = 0$, so $\operatorname{img}(f_{i+1} - s_i d_{i+1}) \subset \ker\partial_{i+1} = \operatorname{img}\partial_{i+2}$, i.e. we have the following commutative diagram:

$$\begin{array}{ccccc} & & X_{i+1} & & \\ & & \big\downarrow{\scriptstyle f_{i+1} - s_i d_{i+1}} & & \\ Y_{i+2} & \xrightarrow{\partial_{i+2}} & \operatorname{img}(\partial_{i+2}) & \longrightarrow & 0. \end{array}$$

Since X_{i+1} is projective, $\exists s_{i+1} : X_{i+1} \to Y_{i+2}$ s.t.

$$\partial_{i+2} s_{i+1} = f_{i+1} - s_i d_{i+1}.$$

□

Now we return to the proof of the theorem. If $F : M \to N$, and f and f' are lifts of F, then $f - f'$ lifts the zero map: $M \to N$. If we take $T \equiv 0$ in the lemma, we get the desired homotopy. □

We want to state an important corollary to this theorem. The corollary will tell us how to compute cohomology, but first we have to say what cohomology is in this setting.

Definition 3.5: Let X be a complex and A a module. Define a sequence of modules and homomorphisms

$$\operatorname{Hom}(M, A) \xrightarrow{\epsilon^*} \operatorname{Hom}(X_0, A) \xrightarrow{\delta^0} \operatorname{Hom}(X_1, A) \xrightarrow{\delta^1} \operatorname{Hom}(X_2, A) \xrightarrow{\delta^2} \ldots \quad (14)$$

by

$$[\epsilon^*(c)](x_0) = c(\epsilon(x_0))$$

for $c \in \mathrm{Hom}(M, A)$ and $x_0 \in X_0$, and

$$[\delta^n(c)](x_{n+1}) = c(dx_{n+1})$$

for $c \in \mathrm{Hom}(X_n, A)$ and $x_{n+1} \in X_{n+1}$. We denote the sequence (14) by $\mathrm{Hom}(X, A)$. If Y is another complex and $f \in \mathrm{Hom}(X, Y)$, then we define $\bar{f}$: $\mathrm{Hom}(Y, A) \to \mathrm{Hom}(X, A)$ by $[\bar{f}(c)](x_n) = c(f(x_n))$ for $c \in \mathrm{Hom}(Y_n, A)$, and $x_n \in X_n$. We define the *cohomology* of $\mathrm{Hom}(X, A)$ by

$$H^0(\mathrm{Hom}(X, A)) = \ker(\delta^0)$$
$$H^n(\mathrm{Hom}(X, A)) = \ker(\delta^n)/\mathrm{img}(\delta^{n-1}) \text{ for } n > 0.$$

Exercise 3.3: i) Show ϵ^* is injective.

ii) If X is a resolution, $H_n(X) = 0$ for $n > 0$. Show $H^n(\mathrm{Hom}(X, A))$ may not be 0.

iii) Show $\overline{f \circ g} = \bar{g} \circ \bar{f}$.

iv) Show $\bar{f}$ induces a map $f^* : H^n(\mathrm{Hom}(Y, A)) \to H^n(\mathrm{Hom}(X, A))$ and $(f \circ g)^* = g^* \circ f^*$.

v) Show that if f and g are homotopic, $f^* = g^*$.

vi) Show that a map $F : A \to B$ induces a map $\underline{F} : \mathrm{Hom}(X, A) \to \mathrm{Hom}(X, B)$ and a map $F_\flat : H^n(\mathrm{Hom}(X, A)) \to H^n(\mathrm{Hom}(X, B))$.

Now we can state our corollary.

Corollary 3.1: *Let X and Y be projective resolutions of M and A any module. Then*

$$H^n(\mathrm{Hom}(X, A)) \approx H^n(\mathrm{Hom}(X, B))$$

Proof: Let $f : X \to Y$ and $g : Y \to X$ both lift the identity map of M. Thus both $f \circ g$ and $g \circ f$ are homotopic to the appropriate identities. Hence both $g^* \circ f^*$ and $f^* \circ g^*$ are the appropriate identities on cohomology, so f^* and g^* must be isomorphisms. □

So we see that the "reasonable" complexes mentioned above are the projective resolutions. We still haven't seen how all these matters relate to $H^2(\Phi; M)$, but we will. Before we do, the formalism of our constructions allows us the following wonderful and useful result:

Theorem 3.2 : *Let*

$$0 \longrightarrow A \xrightarrow{F} B \xrightarrow{G} C \longrightarrow 0$$

be an exact sequence of modules. Then we can define maps

$$\beta^n : H^n(\mathrm{Hom}(X,C)) \to H^{n+1}(\mathrm{Hom}(X,A))$$

so that the following long sequence is exact:

$$\ldots \xrightarrow{\beta^{n-1}} H^n(\mathrm{Hom}(X,A)) \xrightarrow{F_\flat} H^n(\mathrm{Hom}(X,B)) \xrightarrow{G_\flat}$$
$$H^n(\mathrm{Hom}(X,C)) \xrightarrow{\beta^n} H^{n+1}(\mathrm{Hom}(X,A)) \xrightarrow{F_\flat} \ldots$$

Exercise 3.4: Prove this theorem.

The proof can be found in many books (e.g. [68], page 57), but it is worthwhile to try to prove it yourself. It is a complicated "diagram chase." Here is part of the diagram.

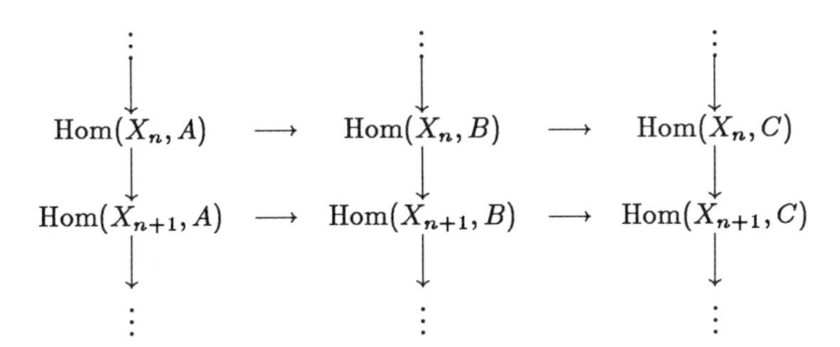

If you can do this diagram chase, you shouldn't be afraid of any other diagram chase.

Definition 3.6: $\beta = \{\beta_n\}$ is called the *connecting homomorphism* or *Bockstein.*

Now we can give a new, more general definition of the cohomology of groups.

Definition 3.7: Let Φ be a group and A a Φ-module. Let X be any projective resolution of $\mathbb{Z}$ (regarded as a trivial Φ-module) for Φ. Then we define the *nth cohomology group of* Φ *with coefficients in* A by

$$H^n(\Phi; A) = H^n(\mathrm{Hom}(X,A)). \tag{15}$$

If you know about categories, you can see that vi) of Exercise 3.3 shows that $H^n(\Phi;\ )$ is a covariant functor. Since Φ appears very indirectly in equation (15), you won't be surprised to learn that H^n is not a functor in the first variable. Now we must show that this new $H^2(\Phi; M)$ is the same as the old one. The trick is to pick the right projective resolution of $\mathbb{Z}$ for Φ, and we do just that in the first example of the next section.

4. Examples

A. *The Standard Non-homogeneous or Bar Resolution*

Let X_n be the free (Φ-)module generated by the set $\{[x_1, \ldots, x_n] : x_i \in \Phi, x_i \neq 1, i = 1, 2, \ldots, n\}$. If $x \in \Phi$, we simply denote the result of x acting on $[x_1, \ldots, x_n]$ as $x[x_1, \ldots, x_n]$. We can think of X_n as the free abelian group generated by $\{x[x_1, \ldots x_n]\}$. For convenience, we can say $[x_1, \ldots, x_n] = 0$ if any $x_i = 1$. Let X_0 be the free module generated by the empty symbol $[\,]$. X_0 is isomorphic to the group ring $\mathbb{Z}[\Phi]$. Define $\epsilon : X_0 \to \mathbb{Z}$ by $\epsilon([\,]) = 1$. Define $\partial_n : X_n \to X_{n-1}$ for $n > 0$ by

$$\begin{aligned}\partial_n[x_1, \ldots, x_n] &= x_1[x_2, \ldots, x_n] \\ &+ \sum_{i=1}^{n-1}(-1)^i[x_1, \ldots, x_i x_{i+1}, \ldots, x_n] + (-1)^n[x_1, \ldots, x_{n-1}]. \end{aligned} \tag{16}$$

So, for example, $\partial_1[x] = x[\,] - [\,]$, and $\partial_2[x, y] = x[y] - [xy] + [x]$.

Exercise 4.1: Show equation (16) makes sense even if some $x_i = 1$.

Theorem 4.1 : *For any group Φ, X with ∂ and ϵ, is a free resolution of $\mathbb{Z}$ for Φ.*

Proof: Clearly each X_n is free, so it suffices to show X is a resolution, i.e. to show

$$\ldots \longrightarrow X_n \xrightarrow{\partial_n} X_{n-1} \longrightarrow \ldots X_1 \xrightarrow{\partial_1} X_0 \xrightarrow{\epsilon} \mathbb{Z} \longrightarrow 0 \tag{17}$$

is exact.

Define $s_{-1} : \mathbb{Z} \to X_0$ by $s_{-1}(1) = [\,]$, and $s_n : X_n \to X_{n+1}$ by $s_n(x[x_1, \ldots, x_n]) = [x, x_1, \ldots, x_n]$. Notice that since we have defined s_n not on a Φ-basis of X_n, but only on a $\mathbb{Z}$-basis, we cannot expect s_n to be a Φ-homomorphism. It is merely an abelian group homomorphism. However, if we can show that (17) is exact as a sequence of abelian groups, it will be exact as a sequence of modules (over Φ). (This is a neat trick that is well worth remembering.) If we can show

$$\epsilon s_{-1} = \text{identity of } \mathbb{Z}, \text{ and} \tag{18}$$

$$\partial_1 s_0 + s_{-1}\epsilon = \text{identity of } X_0, \text{ and} \tag{19}$$

$$\partial_{n+1}s_n + s_{n-1}\partial_n = \text{identity of } X_n, \text{ for } n > 0 \tag{20}$$

we will have shown that the identity maps of X are homotopic to the zero map. If we knew X were a complex, we could conclude it had no homology, and that (17) was exact. For that we need to know that $\partial^2 = 0$. The way to do this is to work out from the middle of the proof. First

Exercise 4.2: Prove that (18), (19), and (20) hold.

Now we use this exercise to show that $\partial^2 = 0$. First we observe that the s_n's togehther with (18), (19), and (20) determine ∂_n, since X_{n+1} is generated as a module by $s_n(X_n)$ and on $s_n(X_n)$, we have

$$\partial_{n+1}s_n = (\text{identity}) - s_{n-1}\partial_n.$$

∂_{n+1} can be determined by induction, i.e. (16) can be deduced from the definition of s_n and (18), (19), and (20). Therefore we know

$$\begin{aligned}\epsilon\partial_1 s_0 &= \epsilon - \epsilon s_{-1}\epsilon = 0, \text{ and}\\ \partial_n\partial_{n+1}s_n &= \partial_n(\text{identity} - s_{n-1}\partial_n)\\ &= \partial_n - (\partial_n s_{n-1})\partial_n\\ &= \partial_n - \partial_n - s_{n-2}\partial_{n-1}\partial_n\end{aligned}$$

which is 0 by induction. Hence $\partial^2 = 0$, and X is a complex , and then the homotopy s shows $H^n(X) = 0 \quad \forall n$, so (16) is exact. □

Exercise 4.3: Show that even if we allow generators to have the form $[x_1, \ldots, x_n]$ with some x_i's$= 1$ (i.e. we do not require $[x_1, \ldots, x_n] = 0$ if some $x_i = 1$), we still get a projective resolution of $\mathbb{Z}$ for Φ.

Now if M is any module, $f \in \operatorname{Hom}(X_n, M)$ is completely determined by its values on a Φ-basis of X, for example, $\{[x_1, \ldots, x_n]\}$, and these values can be arbitrary. Such an f is therefore equivalent to a map $f : \Phi^n \to M$ where Φ^n is the product of Φ with itself n times, e.g. $\Phi^2 = \Phi \times \Phi$. It is now easy to see that the coboundary maps δ^1 and δ^2 used previously are the ones defined in Definition 3.5 for this complex, i.e. (16) yields equations (27) and (28) of Chapter I. So what we have done on extensions fits nicely into the framework of cohomology of groups. If we really want to compute $H^2(\Phi; M)$, we should try to find some smaller resolution of $H^2(\Phi; M)$.

B. *The Standard Homogeneous Resolution*

This resolution is really the same as the previous one; they are isomorphic. Let $\tilde{X}_n$ be the free abelian group generated by $\{(y_0, y_1, \ldots, y_n) : y_i \in \Phi, i = 0, 1, \ldots, n\}$ modulo the relation $(y_0, y_1, \ldots, y_n) = 0$ if $y_i = y_{i-1}$ for $i = 1, 2, \ldots, n$. Make $\tilde{X}_n$ into a module (over Φ) by setting $y \cdot (y_0, \ldots, y_n) = (yy_0, \ldots, yy_n)$. Define $\tilde{\partial}_n : \tilde{X}_n \to \tilde{X}_{n-1}$ for $n > 0$ by

$$\tilde{\partial}_n(y_0, \ldots, y_n) = \sum_{i=0}^{n} (-1)^i (y_0, \ldots, \hat{y}_i, \ldots, y_n) \tag{21}$$

where $\hat{y}_i$ means to omit y_i. $\tilde{X}_0$ is clearly $\mathbb{Z}[\Phi]$, and we define $\tilde{\epsilon} : \tilde{X}_0 \to \mathbb{Z}$ by $\tilde{\epsilon}(\sigma) = 1 \quad \forall \sigma \in \Phi$.

Exercise 4.4: Show that X and $\tilde{X}$ are isomorphic complexes. (**Hint:** The isomorphisms are

$$(y_0, \ldots, y_n) \mapsto y_0[y_0^{-1}y_1, y_1^{-1}y_2, \ldots, y_{n-1}^{-1}y_n], \text{ and}$$
$$x[x_1, \ldots, x_n] \mapsto (x, xx_1, xx_1x_2, \ldots, xx_1x_2\cdots x_n).)$$

The main reason one would use $\tilde{X}$ rather than X is that the boundary formula (21) is somewhat nicer than the boundary formula (16).

C. A Good Resolution for Cyclic Groups

While the resolutions X and $\tilde{X}$ work for any group, they are so big that it is almost impossible to use them to compute cohomology. The main point of the construction of complexes is to be able to pick good resolutions for particular groups. This example is a resolution of $\mathbb{Z}$ for cyclic groups and will allow us to easily prove proposition 5.1 of Chapter I.

Let G be $\mathbb{Z}_n$, the cyclic group of order n. Let Σ and $\Delta \in \mathbb{Z}[G]$ be defined by

$$\Sigma = \sum_{i=0}^{n-1} g^i,$$

and

$$\Delta = g - 1$$

where g is a fixed generator of G.

Notice that we can think of $\mathbb{Z}[G]$ as the polynomial ring $\mathbb{Z}[x]$ modulo the relation $x^n = 0$. We can also think of Σ and Δ as inducing automorphisms of $\mathbb{Z}[G]$ by multiplication.

Proposition 4.1: *i)* $\ker\Delta = \operatorname{img}\Sigma$, *and*
ii) $\operatorname{img}\Delta = \ker\Sigma$.

Proof: It is trivial to see that $\Sigma\Delta = \Delta\Sigma = 0$, so $\operatorname{img}\Delta \subset \ker\Sigma$ and $\operatorname{img}\Sigma \subset \ker\Delta$. Now suppose $u = \sum_{i=0}^{n-1} a_i g^i \in \mathbb{Z}[G]$. Then

$$\Sigma \cdot u = \sum_{i,j=0}^{n-1} a_i g^j \text{ and } \Delta u = \sum_{i=0}^{n} (a_{i-1} - a_i) g^i,$$

where we have put $a_n = a_0$ in the second equation. Therefore $\Delta u = 0 \Rightarrow a_0 = a_1 = \cdots = a_{n-1}$, so $u = \Sigma \cdot a_0$, i.e. $u \in \operatorname{img}\Sigma$, so $\ker\Delta \subset \operatorname{img}\Sigma$.

Suppose $\Sigma \cdot u = 0$. Then we must have $\sum_{i=0}^{n-1} a_i = 0$, so

$$u = \Delta[-a_0 - (a_1 + a_0)g - (a_2 + a_1 + a_0)g^2 + \cdots + (a_{n-1} + a_{n-2} + \cdots + a_1 + a_0)g^{n-1}],$$

which shows that $u \in \operatorname{img}\Delta$, so $\ker\Sigma \subset \operatorname{img}\Delta$. $\square$

Now $\mathbb{Z}[G]$ is obviously a free module over Φ. Therefore the sequence

$$\cdots \longrightarrow \mathbb{Z}[G] \xrightarrow{\Sigma} \mathbb{Z}[G] \xrightarrow{\Delta} \mathbb{Z}[G] \xrightarrow{\Sigma} \mathbb{Z}[G] \xrightarrow{\Delta} \mathbb{Z}[G] \xrightarrow{\Sigma} \mathbb{Z} \longrightarrow 0$$

is a free resolution, which we call W, of $\mathbb{Z}$ for G, i.e. $W_i = \mathbb{Z}[G] \quad \forall i$ and $d_i = \Delta$ for i odd, and $d_i = \Sigma$ for i even.

Proposition 4.2: *Let M be any G-module. Then*

$$H^0(G;M) \approx M^G \overset{\text{def}}{=} \{m \in M : g \cdot m = m\} = \ker\Delta_M,$$
$$H^{2n}(G;M) \approx M^G/\Sigma_M \cdot M \quad (n > 0),$$

and

$$H^{2n+1}(G;M) \approx (\ker\Sigma_M) \cdot M/\Delta_M \cdot M.$$

where Δ_M and Σ_M are the maps induced on M by multiplication by Δ and Σ.

Exercise 4.5: Prove this proposition.

Proposition 5.1 of Chapter I is an easy consequence of this proposition.

D. *A Good Resolution for Free Abelian Groups*

For the cohomological study of Bieberbach groups we hope will be given in Volume II it is important to have a good resolution of the free abelian subgroup of the Bieberbach group. We give one in very explicit form. For a more elegant treatment, see page 188 of [57].

There is a notational difficulty here which turns out to be rather bothersome. We would like to think of $\mathbb{Z}^n$ as an *additive* group, but we must consider $\mathbb{Z}[\mathbb{Z}^n]$, and the formal addition in $\mathbb{Z}[\mathbb{Z}^n]$ is quite different from the group operation in $\mathbb{Z}^n$. We could try to indicate one by "+", and the other by "$\dot{+}$", or "$\oplus$", or something else. This turns out to be a mess. The alternative is to write $\mathbb{Z}^n$ multiplicatively. This keeps everything straight, but produces formulas that look very strange. Our advice is that if you see a bizarre looking formula, rewrite it using additive notation in $\mathbb{Z}^n$, but be careful to distinguish the two different kinds of addition.

Let M be a free abelian group of rank d written multiplicatively. Let $e_1, \ldots, e_d$ be a basis for M. A typical element, m, of M can be written $m = e_1^{a_1} e_2^{a_2} \cdots e_d^{a_d}$ with $a_i \in \mathbb{Z}$ for $i = 1, 2, \ldots, d$. Now $\mathbb{Z}[M]$ is a free M-module of rank 1, but as an abelian group (i.e. a $\mathbb{Z}$-module), it is free of infinite rank. Nevertheless, M and $\mathbb{Z}[M]$ are $\mathbb{Z}$-modules, and we can form the exterior power, $\bigwedge^i M$, as a $\mathbb{Z}$-module (it will be a free $\mathbb{Z}$-module of rank $\binom{d}{i}$)), and then take its tensor product with $\mathbb{Z}[M]$ regarding both as $\mathbb{Z}$-modules. so we get a large free $\mathbb{Z}$-module, $P_i = \bigwedge^i M \otimes \mathbb{Z}[M]$.

We can make P_i into an M-module by setting

$$m \cdot (e_{j_1} \wedge e_{j_2} \wedge \cdots \wedge e_{j_i} \otimes x) = me_{j_1} \wedge me_{j_2} \wedge \cdots \wedge me_{j_i} \otimes mx \tag{22}$$

for $m \in M$ and $x \in \mathbb{Z}[M]$ (recall M is being written multiplicatively). This kind of action, where the element acts on some kind of product by acting on each factor of the product, is sometimes called a *diagonal action.* We want to show P_i is M-free, i.e. free as a module over M. There are two ways to do this. We do one and leave the other as an exercise.

Lemma 4.1: *Let N be a module over M which is finitely generated as an abelian group, and let $P = n \otimes \mathbb{Z}[M]$ as an abelian group. Make P into a module over M by the diagonal action, i.e.*

$$m(n \otimes m') = mn \otimes mm' \text{ for } m, m' \in M \text{ and } n \in N.$$

Then P is M-free.

Proof: By hypothesis, we can find a $\mathbb{Z}$-basis, say $n_1, \ldots n_k$, for N. Let P' be the M-module $N \otimes \mathbb{Z}[M]$, but with the action

$$m \star (n \otimes m') = n \otimes mm' \text{ for } m, m' \in M \text{ and } n \in N. \tag{23}$$

Then it is easy to see that P' is free as an M-module, i.e. $\{n_i \otimes 1\}$ is a basis for P' over M.

Define a map $f : P \to P'$ by

$$f(n_i \otimes m) = m^{-1} \cdot n_i \otimes m.$$

Now

$$\begin{aligned} f(m' \cdot (n_i \otimes m)) &= f(m'n_i \otimes m'm) \\ &= m^{-1}(m')^{-1}m'n_i \otimes m'm \\ &= m^{-1} \cdot n_i \otimes m'm, \end{aligned}$$

while

$$m' \star f(n_i \otimes m) = m' \star (m^{-1} \cdot n_i \otimes m) = m^{-1} \cdot n_i \otimes m'm$$

by (23). Hence f is a homomorphism of M-modules. It is easy to see that f is an isomorphism (i.e. bijective). □

Exercise 4.6: Let N be the free module over M of rank d (which is the rank of M itself over $\mathbb{Z}$). Show P_i is isomorphic to $\bigwedge_M^i N$ where now the exterior power is taken over M (i.e. over $\mathbb{Z}[M]$). Hence P_i is a free M-module.

In dealing with P_i, we often have the choice of defining a map on a $\mathbb{Z}$-basis and then showing it is a M-homomorphism, or defining it on an M-basis, but then we must usually show it is well-defined.

Let P be the complex defined by $P = \{P_i\}$ together with a boundary map $\partial_i : P_i \to P_{i-1}$ for $i > 0$ defined by

$$\partial_i(e_{j_1}\wedge e_{j_2}\wedge\cdots\wedge e_{j_i}\otimes x) = \sum_{k=1}^{i}(-1)^k(e_{j_1}\wedge e_{j_2}\wedge\cdots\wedge \hat{e}_{j_k}\wedge\cdots\wedge e_{j_i}\otimes(1-e_{j_k})\cdot x) \tag{24}$$

where $e_1, \ldots, e_d$ is a $\mathbb{Z}$-basis of M and $x \in \mathbb{Z}[M]$.

Exercise 4.7: Show $\partial_i\partial_{i-1} = 0$.

Now $P_0 = \bigwedge^0 M \otimes \mathbb{Z}[M] = \mathbb{Z} \otimes \mathbb{Z}[M] \approx \mathbb{Z}[M]$, so we can define an augmentation $\epsilon : P_0 \to \mathbb{Z}$ as usual, i.e. $\epsilon(m) = 1 \quad \forall m \in M$.

Exercise 4.8: Show $H^k(P) = 0$ for $i > 0$.

Hence P (together with ∂ and ϵ) is a free resolution of $\mathbb{Z}$ for M.

Remarks: i) We prefer the description $P_i = \bigwedge^i M \otimes \mathbb{Z}[M]$ to the one of Exercise 4.6, $P_i = \bigwedge_M^i N$, because it naturally breaks the $\mathbb{Z}$-basis, $\{e_{j_1} \wedge \cdots \wedge e_{j_i} \otimes m\}$, of P into two factors. The first, $e_{j_1} \wedge \cdots \wedge e_{j_i}$, contains the important boundary information, while the second, m, contains the important M-action information.

ii) There is a geometric way of thinking about $\{e_{j_1} \wedge \cdots \wedge e_{j_i} \otimes m\}$. Think of M as the integral lattice in $\mathbb{R}^n$. Then $\{e_{j_1}\wedge\cdots\wedge e_{j_i}\otimes m\}$ represents an i-dimensional face of a cell of that lattice. The cell is the one named by m (say m is the "lower left hand" vertex or some other convention), and the face is the one corresponding to the face $\{e_{j_1} \wedge \cdots \wedge e_{j_i}\}$ in the basic cell. For example, for $n = 2$, we have

Proposition 4.3: *If M is a free abelian group of rank d and A is any M-module, then*

i) $H^p(M;A) = 0$ if $p > d$, and

ii) if M acts trivially on A, then $H^p(M;A)$ is isomorphic to the direct sum of $\binom{p}{d}$ copies of A.

Exercise 4.9: Prove this proposition. (**Hint:** For ii), note that the coboundary map in $Hom(P, A)$ is identically 0.)

E. *Spaces of Type* $K(\pi, 1)$

This example is a digression within a digression.

Definition 4.1: Let π be any group. A connected topological space X is said to be a $K(\pi,1)$ or *a space of type* $K(\pi,1)$ if $\pi_1(X) \approx \pi$ and $\pi_i(X) = 0$ for $i > 1$.

A complete locally affine (manifold with a symmetric flat connection) is a $K(\pi,1)$ by Corollary 3.3 of Chapter I. A compact flat manifold is a $K(\pi,1)$, and π must be a Bieberbach group.

If X is any $K(\pi,1)$, then $\tilde{X}$, the universal covering space of X, will satisfy $\pi_i(X) = 0$ for $i > 0$. If X is at all reasonable, $\tilde{X}$ will be contractible, and we will also have $H^i(X;C) = 0$ for $i > 0$, for all coefficients C and for all reasonable cohomology theories. Note that $H^i(X) \neq 0$ in general, but we can say pretty much what $H^i(X)$ is. Let $S(\tilde{X})$ be the chain complex of $\tilde{X}$ for some reasonable cohomology theory, say singular theory. So in this case, $S_0(\tilde{X})$ is the free abelian group generated by all continuous maps from a fixed point p_0 to $\tilde{X}$. Fix a basepoint $\tilde{x} \in \tilde{X}$, and define a homomorphism $\epsilon : S_0(\tilde{X}) \to \mathbb{Z}$ by $\epsilon(s) = 0$ unless $s(p_0) = \tilde{x}$, and $\epsilon(s) = 1$ in this case. So we get a sequence

$$\cdots \longrightarrow S_n(\tilde{X}) \xrightarrow{\partial_n} S_{n-1}(\tilde{X}) \longrightarrow \cdots \longrightarrow S_1(\tilde{X}) \xrightarrow{\partial_1} S_0(\tilde{X}) \xrightarrow{\epsilon} \mathbb{Z} \longrightarrow 0.$$

This sequence is exact since $H_i(\tilde{X}) = H_i(S(\tilde{X})) = 0$ for $i > 0$. (You can worry about what happens at $S_0(\tilde{X})$.) Now $\pi = \pi_1(X)$ acts on $\tilde{X}$ by covering (or deck) transformations, so it also acts on $S_i(\tilde{X})$ by $(\alpha \cdot s)(p) = \alpha \cdot s(p)$ for $\alpha \in \pi$, $s \in S_i(\tilde{X})$, and p in the standard i-simplex. Furthermore, this action is free since the action of π on $\tilde{X}$ is free. Therefore, each $S_i(\tilde{X})$ is a free π-module, so $S(\tilde{X})$ is a free resolution of $\mathbb{Z}$ for π, and we have proved the following:

Proposition 4.4: *Let X be a space of type $K(\pi,1)$ and A be a trivial π-module. Then $H^i(X;A) \approx H^i(\pi;A)$.*

The proposition is still true if the action of π on A is not trivial, but then $H^i(X;A)$ must be interpreted as cohomology with local coefficients.

These results show us that the cohomology of a flat manifold is the same as the cohomology of its fundamental group. You might wonder for which groups besides Bieberbach groups can you find spaces of type $K(\pi,1)$. If you look in [59], you will see that you can always get nice spaces of type $K(\pi,1)$ for any group π.

We can use the ideas of this example to prove two famous theorems.

Theorem 4.2 (P. A. Smith): *Let π be a group which acts freely on $\mathbb{R}^n$. Then π is torsionfree.*

Proof: If π acts freely on $\mathbb{R}^n$, then any cyclic subgroup, ρ, does too. If π is not torsionfree, we can choose ρ to be finite. Then $\mathbb{R}^n/\pi$ will be an n-manifold, so

$$H^i(\mathbb{R}^n/\rho;\mathbb{Z}) = 0 \text{ for } i > n.$$

But $\mathbb{R}^n/\rho$ is a $K(\rho,1)$, so

$$H^i(\mathbb{R}^n/\rho;\mathbb{Z}) \approx H^i(\rho;\mathbb{Z}),$$

and Proposition 3.2 shows this is not 0 for all $i > n$. □

Corollary 4.1: *No finite group can act freely on $\mathbb{R}^n$.*

Definition 4.2: We say a group π has a period p if

$$H^{i+p}(\pi;\mathbb{Z}) \approx H^i(\pi;\mathbb{Z}) \quad \forall i.$$

Theorem 4.3 (Artin and Tate): *If a finite group acts freely on S^p, then it has period $p+1$.*

Proof: Let $C_n = S_n(S^p)$, the n-dimensional singular chains of S^p. Now we know $H_i(S^p;\mathbb{Z}) = 0$ unless $i = 0$ or p, and $H_0(S^p;\mathbb{Z}) \approx H_p(S^p;\mathbb{Z}) \approx \mathbb{Z}$. Hence the sequence

$$0 \longrightarrow \partial_1 C_1 \longrightarrow C_0 \xrightarrow{\epsilon} \mathbb{Z} \longrightarrow 0$$

is exact, and then the sequence

$$C_2 \xrightarrow{\partial_2} C_1 \xrightarrow{\partial_1} C_0 \xrightarrow{\epsilon} \mathbb{Z} \longrightarrow 0$$

is exact. In addition, the sequence

$$0 \longrightarrow \partial_{p+1} C_{p+1} \longrightarrow \ker \partial_p \longrightarrow \mathbb{Z} \longrightarrow 0$$

is exact (this just says that $H_p(S^p; \mathbb{Z}) \approx \mathbb{Z}$). Now let i be the inclusion of $\ker\partial_p$ into C_p. Then we get a free resolution of $\mathbb{Z}$ for π by "splicing" sequences of length p together as follows:

$$\cdots \xrightarrow{\partial_3} C_2 \xrightarrow{\partial_2} C_1 \xrightarrow{\partial_1} C_0 \xrightarrow{0 \oplus \epsilon} \ker\partial_p \oplus \mathbb{Z} \xrightarrow{i \oplus 0} C_p \xrightarrow{\partial_p} C_{p-1}$$

$$\longrightarrow \cdots \longrightarrow C_2 \longrightarrow C_1 \longrightarrow C_0 \longrightarrow \mathbb{Z} \longrightarrow 0,$$

and it is easily seen from this that π has period p. □

Remarks: i) If π is not necessarily finite, but acts freely and properly on S^p, one can show it *is* finite and has period $p+1$.

ii) The more elegant treatment of the cohomology of a free abelian group alluded to above uses a formula for the cohomology of $G \times H$ in terms of the cohomology of G and the cohomology of H. This formula is called the Künneth Formula. You can also use it to conclude that $\mathbb{Z}_p \times \mathbb{Z}_p$ has no periods at all. Hence you can prove that if a finite group acts freely on S^p, all of its Sylow subgroups must be cyclic. You might wonder which groups have this property, and can they all act freely on some sphere. See [60] for some answers.

5. Holonomy Groups

We are now in a position to begin proving that any finite group Φ is the holonomy group of a Bieberbach group and hence of a flat manifold. We need two standard constructions for this proof.

Definition 5.1: Let Φ be a finite group, ρ a subgroup of Φ, and A and C Φ-modules. Define a map $t : \mathrm{Hom}_\rho(C, A) \longrightarrow \mathrm{Hom}_\Phi(C, A)$ by

$$(t(f))(c) = \sum_i x_i \cdot f(x_i^{-1} \cdot c)$$

for $f \in \mathrm{Hom}_\rho(C, A)$, and $c \in C$, and where $\{x_1\rho, \ldots, x_r\rho\}$ is the set of all distinct cosets of ρ in Φ (so $r = |\Phi/\rho|$). We call t the *transfer* (from ρ to Φ).

To see that t is well-defined, replace x_i by $x_i y$ for some $y \in \rho$. Then $x_i y f(y^{-1} x_i^{-1} \cdot c) = x_i f(x_i^{-1} \cdot c)$ since $f \in \mathrm{Hom}_\rho(C, A)$. We must also see that $t(f) \in \mathrm{Hom}_\Phi(C, A)$. Let $x \in \Phi$. Then

$$(t(f))\,(x \cdot c) = \sum_i x(x^{-1}x_i) \cdot f((x^{-1}x_i)^{-1} \cdot c) = x \cdot [(t(f))\,(c)]$$

since $\{x^{-1}x_1, \ldots, x^{-1}x_r\}$ is also a set of coset representatives for ρ in Φ.

Remark: Since any Φ-map is a ρ-map, there is a natural map from $\mathrm{Hom}_\Phi(C, A)$ to $\mathrm{Hom}_\rho(C, A)$. The transfer shows us how to map the other way. We would like to get the transfer by averaging over the cosets of ρ, but the best we can do is to merely add up since we can't divide by r. Therefore, we can expect some factors of r to appear in formulas involving the transfer (and they do).

Now let X be a projective resolution of $\mathbb{Z}$ for Φ.

Exercise 5.1: i) Check that X is a projective resolution of $\mathbb{Z}$ for ρ.

ii) Show that the transfer map $t : \mathrm{Hom}_\rho(X_i, A) \to \mathrm{Hom}_\Phi(X_i, A)$ is a cochain map, i.e. $\delta \circ t = t \circ \delta$.

Definition 5.2: By the exercise above, t induces a map: $H^i(\rho; A) \to H^i(\Phi; A)$ which by an abuse of terminology, we again denote by t, and call the *transfer*.

Since $\mathrm{Hom}_\Phi(X_i, A) \subset \mathrm{Hom}_\rho(X_i, A)$, we have a map $R : H^i(\Phi; A) \to H^i(\rho; A)$. R is called the *restriction*.

Proposition 5.1: *Let $a \in H^i(\Phi; A)$. Then $t \circ R(a) = ra = |\Phi/\rho| \cdot a$.*

Proof: If we consider the transfer on $\mathrm{Hom}_\rho(X_i, A)$, we see that if $f \in \mathrm{Hom}_\Phi(X_i, A)$, then

$$(t(f))\,(x) = \sum_i x_i \cdot f(x_i^{-1} \cdot x) = \sum_i f(x) = r \cdot x$$

which is exactly what we want. $\square$

Before we go on, we would like to record some important facts about the transfer in the following:

Theorem 5.1 : *Suppose ρ is a p-Sylow subgroup of a finite group Φ and A is any Φ-module, then t and R satisfy*

i) $t(H^*(\rho; A))$ *is the p-primary component of* $H^*(\Phi; A)$,
ii) $R\,|$*(p-primary component of* $H^*(\Phi; A)$*) is a monomorphism, and*
iii) $H^*(\rho; A) = \operatorname{img} R \oplus \ker t$.

A proof may be found on page 259 of [16].

Definition 5.3: Let M be a ρ-module. $\mathbb{Z}[\Phi]$ is also a ρ-module which we think of as a right ρ-module. Define the *induced* module by

$$\mathcal{I}^{\Phi}(M) = \mathbb{Z}[\Phi] \otimes_\rho M.$$

It is a left Φ-module where we let Φ act only on $\mathbb{Z}[\Phi]$, i.e. $x \cdot (y \otimes m) = xy \otimes m)$ for $x, y \in \Phi$ and $m \in M$.

We want to examine this construction in a more explicit fashion. Suppose $\Phi = x_1\rho \cup x_2\rho \cup \cdots \cup x_r\rho$ is a coset decomposition of ρ, and say $x_1 = 1$, so any $x \in \Phi$ can be uniquely written $x = x_i a$ for some $a \in \rho$. Therefore, every element in $\mathbb{Z}[\Phi]$ can be written as $\sum_i x_i b_i$ with $b_i \in \mathbb{Z}[\rho]$ for $i = 1, 2, \ldots, r$, i.e.

$$\mathbb{Z}[\Phi] = x_1\mathbb{Z}[\rho] \oplus \cdots \oplus x_r\mathbb{Z}[\rho]$$

which implies that $\mathbb{Z}[\Phi]$ is a free ρ-module with basis $\{x_1, \ldots, x_r\}$. Therefore,

$$\mathcal{I}^{\Phi}(M) = (x_1\mathbb{Z}[\rho] \otimes_\rho M) \oplus \cdots \oplus (x_r\mathbb{Z}[\rho] \otimes_\rho M). \tag{25}$$

Note that this decomposition is not a decomposition as Φ-modules or even as ρ-modules, but merely as free abelian groups (see page 20 of [68]). However (25) does show that the rank of $\mathcal{I}^{\Phi}(M)$ as a Φ-module is r times the rank of M as a ρ-module. Also if $\{m_1, \ldots, m_k\}$ is a $\mathbb{Z}$-basis for M, then $\{x_i \otimes m_j : 1 \leq i \leq r \text{ and } 1 \leq j \leq k\}$ is a $\mathbb{Z}$-basis for $\mathcal{I}^{\Phi}(M)$.

Let us suppose we have such a $\mathbb{Z}$-basis for M, and for $b \in \rho$, let $T(b)$ be the matrix of the map b induces on M (with respect to the given basis). Then extend the map T to Φ by setting $T(x) = 0$ if $x \notin \rho$. For $x \in \Phi$, let $U(x)$ be the matrix of the map x induces on $\mathcal{I}^{\Phi}(M)$ with respect to the basis $\{x_i \otimes m_j\}$.

Exercise 5.2: Show

$$U(x) = \begin{pmatrix} T(x_1^{-1}xx_1) & \cdots & T(x_1^{-1}xx_r) \\ \vdots & & \vdots \\ T(x_r^{-1}xx_1) & \cdots & T(x_r^{-1}xx_r) \end{pmatrix}.$$

Since for any x, there is a unique i for which $x_i^{-1}xx_j \in \rho$, each row and column of blocks has precisely one non-zero block.

Let's look a little more closely at the decomposition (25). As we remarked, (25) is not a decomposition as ρ-modules since, in general, $x_i\mathbb{Z}[\rho]\otimes_\rho M$ is not a ρ-submodule of $\mathcal{I}^\Phi(M)$, i.e. if $y \in \rho$, $yx_i\mathbb{Z}[\rho] \neq x_i\mathbb{Z}[\rho]$. Rather, it is $x_j\mathbb{Z}[\rho]$ for some j that need not be i. However, if $i = 1$, recall we have chosen $x_1 = 1$, so $yx_1\mathbb{Z}[\rho] = x_1y\mathbb{Z}[\rho] = x_1\mathbb{Z}[\rho]$, i.e. $x_1\mathbb{Z}[\rho] \otimes_\rho M$ is a ρ-submodule of $\mathcal{I}^\Phi(M)$. Hence the map $f : \mathcal{I}^\Phi(M) \to M$ defined by

$$f(x_i \otimes m) = \begin{cases} 0, & i \neq 1; \\ m, & i = 1 \end{cases}$$

is a ρ-homomorphism.

Lemma 5.1: *Let $\alpha \in H^j(\rho; M)$ be arbitrary. Then $\exists$ a class $\beta \in H^j(\Phi; \mathcal{I}^\Phi(M))$ s.t. $f_* \circ R(\beta) = \alpha$.*

Proof: Let $k : M \to \mathcal{I}^\Phi(M)$ be the map $m \mapsto x_1 \otimes m$. Let

$$t : \mathrm{Hom}_\rho(A, \mathcal{I}^\Phi(M)) \to \mathrm{Hom}_\Phi(A, \mathcal{I}^\Phi(M))$$

be the transfer where A is any ρ-module. Consider the composition

$$\begin{aligned} \mathrm{Hom}_\rho(A, M) \xrightarrow{\circ k} \mathrm{Hom}_\rho(A, \mathcal{I}^\Phi(M)) \xrightarrow{t} \mathrm{Hom}_\Phi(A, \mathcal{I}^\Phi(M)) \xrightarrow{R} \\ \mathrm{Hom}_\rho(A, \mathcal{I}^\Phi(M)) \xrightarrow{\circ f} \mathrm{Hom}_\rho(A, M) \end{aligned}$$

where $\circ k$ and $\circ f$ mean to compose with k and f respectively. Suppose this composition, $fRtk$, were the identity. Then for A, we could take X_i, where X is any projective resolution of $\mathbb{Z}$ for Φ (and ρ), and then define $\beta \in H^j(\Phi; \mathcal{I}^\Phi(M))$ by $\beta = t \circ k_*(\alpha)$. Then we would have $f_*R(\beta) = f_*Rtk_*(\alpha) = \alpha$, and we would be done.

Let $c \in \mathrm{Hom}_\rho(A, M)$ and $a \in A$. Then

$$[k(c)] = x_1 \otimes c(a) \in \mathcal{I}^\Phi(M)$$

and

$$\begin{aligned} [tk(c)] &= \sum_i x_i\,[k(c)]\,(x_i^{-1}a) \\ &= \sum_i x_i[x_1 \otimes c(x_i^{-1}a)] \\ &= \sum_i x_i \otimes c(x_i^{-1}a) \end{aligned}$$

since $x_1 = 1$, and the action on $I^\Phi(M)$ is all on the first factor. Hence

$$[fRtk(c)](a) = \sum_i f(x_i \otimes c(x_i^{-1}a)) = c(x_1^{-1}a) = c(a)$$

and we are done. □

Now we can finally prove the Auslander-Kuranishi Theorem on holonomy groups of Bieberbach groups (and flat manifolds). The proof we give is due to A. T. Vasquez (see [17]), and only uses the standard machinery developed above. Auslander and Kurnishi's original proof in [6] used some special results of R. Lyndon which we avoid.

Theorem 5.2 (Auslander and Kuranishi): *Let Φ be any finite group. Then there is a Bieberbach group π with $r(\pi) \approx \Phi$ and a flat manifold X with $\Phi(X) \approx \Phi$.*

Proof: By Theorem 2.1, it suffices to find a Φ-module N s.t. $H^2(\Phi; N)$ contains a class α s.t. $R_\rho(\alpha) \neq 0$ where R_ρ is the restriction to any cyclic subgroup ρ of Φ. What we do is to find, for each such ρ, a Φ-module, N_ρ, with $\alpha_\rho \in H^2(\Phi; N)$ s.t. $R_\rho(\alpha) \neq 0$. Then we can define

$$N = \mathbb{Z}[\Phi] \oplus (\oplus_{\rho \subset \Phi} N_\rho) .$$

The sum is finite since Φ is finite. The term $\mathbb{Z}[\Phi]$ is added to insure that the module is faithful. Now by Proposition 5.2 of Chapter I,

$$H^2(\Phi; N) = \oplus_{\rho \in \Phi} H^2(\Phi; N_\rho),$$

since $H^2(\Phi; \mathbb{Z}[\Phi]) = 0$ (why?). So the trick is to get the modules N_ρ. But we can use Lemma 5.1 with $M = \mathbb{Z}$ (with trivial ρ-action) and α arbitrary in $H^2(\Phi; \mathbb{Z}) - \{0\}$. Then we take $I^\Phi(\mathbb{Z})$ for N_ρ, and α_ρ is the class β of Lemma 5.1. If $R(\beta)$ were 0, then $f_* \circ R(\beta)$ would be 0 which we have excluded. □

Chapter IV

Holonomy Groups of Prime Order

1. Introduction

We now give some examples of how to apply the general classification scheme of Chapter III. By Theorem 5.2 of Chapter II, we can pick any group we want for the holonomy group Φ. It is, of course, trivial to see that the only Bieberbach groups with trivial holonomy groups (i.e. $\Phi = \{1\}$) are the free abelian groups, so the only compact riemannian manifolds with trivial holonomy group are the flat tori. Notice that we did not have to say that the riemannian manifold was "flat" since by Theorem 3.2 of Chapter II, any manifold with a finite (or even merely totally disconnected) holonomy group must have zero curvature.

The next group to look at would naturally be $\mathbb{Z}_2$. It turns out that it is actually possible to classify those Bieberbach groups whose holonomy group has prime order. Since this includes the case of $\mathbb{Z}_2$, we will do this next. A basic reference for this chapter is [17].

According to our general scheme, the first thing to do is to look at faithful integral representations of our group Φ. In order to do this, we need to do some algebraic number theory. In this chapter, Φ will always be a group of prime order p.

2. Digression — Some Algebraic Number Theory

We are going to develop the algebraic number theory we need only in a very special case, namely we will do it for only one class of rings of algebraic numbers, namely the ring of cyclotomic integers (see below). This approach will simplify some situations and has the virture of being very explicit. Of course, it goes against the philosophy of doing mathematics in as much generality as possible, but the general theory can be looked up in many places. Chapter III of [28] is particularly appropriate for our purposes.

First, however, we will give some background and motivation. One of the basic theorems of classical number theory is Euclid's theorem that every positive integer can be uniquely expressed (up to order) as a product of prime numbers. This is sometimes called the Fundamental Theorem of Arithmetic. It is a theorem about the ring of integers, $\mathbb{Z}$. One could (and should) wonder about its validity in other rings. Some rings which are closely related to $\mathbb{Z}$ are what could be called *algebraic number rings.* These are the rings obtained from $\mathbb{Z}$ by adjoining some root of a polynomial (with integer coefficients) to $\mathbb{Z}$. For example, you might adjoin $\sqrt{2}$ to $\mathbb{Z}$ to get the ring $\mathbb{Z}[\sqrt{2}]$. $\mathbb{Z}[\sqrt{2}]$ consists of elements of the form $a + b\sqrt{2}$ with $a, b \in \mathbb{Z}$ and

$$(a + b\sqrt{2}) + (a' + b'\sqrt{2}) = (a + a') + (b + b')\sqrt{2}$$

and

$$(a + b\sqrt{2})(a' + b'\sqrt{2}) = (aa' + 2bb') + (ab' + ba')\sqrt{2}.$$

Other examples are $\mathbb{Z}[\sqrt{-21}]$ and the gaussian integers, $\mathbb{Z}[i]$, where $i^2 = -1$. Some of these rings satisfy the Fundamental Theorem, and others don't. In all of them, any element can be written as a (finite) product of prime elements. What may go wrong is the uniqueness. The question is whether these rings are unique factorization domains (UFD's). For example, $\mathbb{Z}[\sqrt{2}]$ is a UFD, while $\mathbb{Z}[\sqrt{-21}]$ is not since $(\sqrt{-21})(\sqrt{-21}) = -3 \cdot -7$.

In order to understand the above situation, Dedekind invented what he called *ideal numbers* which we now call simply *ideals.* Dedekind showed that in a "nice" algebraic number ring, ideals satisfy the Fundamental Theorem, i.e. every ideal can be written uniquely as a product of prime ideals. The rings we are interested in are, of course, "nice". (For those in the know, "nice" means intergrally closed.) Since every element of the ring defines a (principal) ideal, we see that the failure of unique factorization is due to the absence of some "numbers" that "should" be in the ring, but are not. These "missing numbers" correspond to the prime ideals that are not principal.

If all the ideals in an algebraic number ring are principal, we do have unique factorization, but if they are not all principal, we do not. We would like to define an invariant which would tell us "how many" non-principal ideals there are. This is easily done, and turns out to lead to a new branch of number theory, *class field theory.*

Definition 2.1: We say two ideals $\mathfrak{a}$ and $\mathfrak{b}$ in an algebraic number ring

R are *equivalent* if there are elements $x, y \in R$ s.t. $xa = yb$. The set of these equivalence classes is called the *ideal class group* of R and is denoted by $C(R)$. Its order is called the *class number* of R and is denoted by h_R.

You may rightfully object that it is not at all obvious that $C(R)$ is a group. You can multiply ideals, and it is easy to see that this induces an associative multiplication on $C(R)$. It is easy to see that all principal ideals are equivalent, i.e. $x \cdot (y) = y \cdot (x)$ where (x) is the principal ideal generated by x. It is also easy to see that the class of the principal ideals acts as an identity for this multiplication. But what about inverses? Well, they exists, and if we had used somewhat fancier ideas (and ideals), namely fractional ideals, it would be clear that inverses exists, but since we have no need for the group structure of $C(R)$, we won't go into all that. We just call $C(R)$ a group because that is the standard terminology.

It is also not clear that $C(R)$ is finite. This is a well-known non-trivial result which is related to some of the finiteness results of Section 6 of Chapter I. A reference for $C(R)$ is [28] starting on page 125.

It should be clear that h_R measures how much unique factorization fails in R. $C(R)$, itself, is a much more delicate measure of the same thing. In general, h_R, let alone $C(R)$, is very difficult to compute. We will later show that in the case of the ring we are most interested in, h_R is related to one of the deepest and most difficult problems in mathematics (see the remarks at the end of Section 8).

Recall that p is a fixed prime.

Definition 2.2: Let $\varsigma \neq 1$ be a fixed pth root of one. We call the ring $\mathbb{Z}[\varsigma]$ the *cyclotomic ring of pth roots of unity*, and usually denote it by R_p or R. Let K be the quotient field of R, so $K = \mathbb{Q}(\varsigma)$. K is called the *cyclotomic field of pth roots of unity.*

Clearly ς is a root of the polynomial $X^p - 1 = 0$. Now $X^p - 1$ is not irreducible, since

$$X^p - 1 = (X - 1)(X^{p-1} + X^{p-2} + \cdots + X + 1). \tag{1}$$

Is $X^{p-1} + \cdots + X + 1$ irreducible? To find out we use the following celebrated result:

Theorem 2.1 (Eisenstein's Irreducibility Criterion): *Let*

$$f(X) = X^n + a_{n-1}X^{n-1} + \cdots + a_1 X + a_0 \tag{2}$$

be a polynomial in $\mathbb{Z}[X]$. *Suppose* p *is a prime s.t.* $p | a_i$ *for* $i = 0, 1, \ldots, n-1$, *but* $p^2 \not| a_0$. *Then* f *is irreducible as a polynomial in* $\mathbb{Q}[X]$.

A proof can be found on page 42 of [72], for example.

Proposition 2.1: $X^{p-1} + \cdots + X + 1$ *is irreducible over* $\mathbb{Q}$.

Proof: Letting $Y = X - 1$, we have

$$f(X) = X^{p-1} + \cdots + X + 1 = \frac{X^p - 1}{X - 1} = \frac{(Y+1)^p - 1}{Y}$$

$$= Y^{p-1} + \sum_{j=1}^{p-1} \binom{p}{j} Y^{j-1} \stackrel{\text{def}}{=} f_1(Y).$$

If f_1 is irreducible over $\mathbb{Q}$, so is f. Now $p | \binom{p}{j}$ for $j = 1, 2, \ldots, p-1$, but $p^2 \not| \binom{p}{1}$, so the Eisenstein Criterion says f_1 is irreducible over $\mathbb{Q}$. □

$X^{p-1} + X^{p-2} + \cdots + X + 1$ is called the *cyclotomic polynomial.*

Exercise 2.1: Show $K = \mathbb{Q}(\varsigma)$ is a vector space of dimension $p - 1$ over $\mathbb{Q}$ with basis $1, \varsigma, \varsigma^2, \ldots, \varsigma^{p-2}$. Show also that $R = \mathbb{Z}[\varsigma]$ is a free abelian group with the same basis.

We need some tools for working with R and K, and these tools are provided by linear algebra As a reference for linear algebra, you can see [L] for example.

Definition 2.3: Let $x \in K$. Regarding K as a vector space over $\mathbb{Q}$, multiplication by x induces a linear map $l_x : K \to K$. We call the *trace* (respectively *norm*, respectively *characteristic polynomial*, respectively *minimal polynomial*) of x the trace (respectively determinant, respectively characteristic polynomial, respectively minimal polynomial) of l_x. We write $\mathrm{tr}(x)$ and $\mathrm{N}(x)$ for the trace and norm of x respectively. Note that $\mathrm{N}(x)$ and $\mathrm{tr}(x)$ are in $\mathbb{Q}$.

Exercise 2.2: Let $a \in \mathbb{Q}$. Show

i) $\mathrm{tr}(x + x') = \mathrm{tr}(x) + \mathrm{tr}(x')$,
ii) $\mathrm{tr}(ax) = a \cdot \mathrm{tr}(x)$,
iii) $\mathrm{tr}(a) = (p-1)a$,
iv) $\mathrm{N}(xx') = \mathrm{N}(x)\mathrm{N}(x')$,
v) $\mathrm{N}(a) = a^{p-1}$, and
vi) $\mathrm{N}(ax) = a^{p-1}\mathrm{N}(x)$.

We need a tool for computing norms and traces.

Proposition 2.2: *Let $x \in K$. Let F be the minimal polynomial of x and suppose it has roots $\bar{x}_1, \ldots, \bar{x}_r$ in $\mathbb{C}$. Let m be the dimension of K considered as a vector space over the field $\mathbb{Q}(x)$. Let $x_1, \ldots, x_n$ be the roots $\bar{x}_1, \ldots, \bar{x}_r$ each repeated m times (so $n = mr$). Then $n = p - 1$,*

$$\mathrm{tr}(x) = x_1 + \cdots + x_{p-1} \tag{3}$$

$$\mathrm{N}(x) = x_1 \cdots x_{p-1}, \tag{4}$$

and the characteristic polynomial of x is F^m.

Proof: First we do the case of x s.t. $K = \mathbb{Q}(x)$, e.g. $x = \varsigma$ is such a case. In this case $m = 1$ and $r = p - 1$. Consider the map $\Theta : \mathbb{Q}(X) \to K$ defined by $\Theta(a) = a$ if $a \in \mathbb{Q}$ and $\Theta(X) = x$. Since x satisfies its minimal polynomial, $F(x) = 0$ and $F \in \ker\Theta$. Since F is minimal, $\ker\Theta$ is the principal ideal generated by F, i.e. $\ker\Theta = (F)$. Hence K is isomorphic to $\mathbb{Q}(X)/(F)$. Now $1, X, \ldots, X^{p-2}$ is a basis for K over $\mathbb{Q}$. Suppose

$$F(X) = X^{p-1} + a_{p-2}X^{p-2} + \cdots + a_1 X + a_0$$

with $a_i \in \mathbb{Q}$. We can pick the leading coefficient to be 1 since $\mathbb{Q}$ is a field. This is the usual convention for the minimal polynomial. Since $x \cdot x^i = x^{i+1}$ if $i < n$, and $x \cdot x^{p-2} = x^{p-1} = -a_{p-2}x^{p-2} - \cdots - a_1 x - a_0$, the matrix of the linear map l_x w.r.t. the basis $1, x, \ldots, x^{p-2}$ is

$$\begin{pmatrix} 0 & 0 & \cdots & 0 & -a_0 \\ 1 & 0 & \cdots & 0 & -a_1 \\ 0 & 1 & \cdots & 0 & -a_2 \\ \vdots & \vdots & & \vdots & \vdots \\ 0 & 0 & \cdots & 1 & -a_{p-2} \end{pmatrix}.$$

The determinant of $X \cdot I_{n-1} - l_x$ is

$$\begin{vmatrix} X & 0 & \cdots & 0 & a_0 \\ -1 & X & \cdots & 0 & a_1 \\ 0 & -1 & \cdots & 0 & a_2 \\ \vdots & \vdots & & \vdots & \vdots \\ 0 & 0 & \cdots & X & a_{p-3} \\ 0 & 0 & \cdots & -1 & X + a_{p-2} \end{vmatrix}.$$

We expand this as a polynomial in X, and we get the characteristic polynomial of x which in this case is just F, the minimal polynomial. But we know from linear algebra that the characteristic polynomial satisfies

$$\det(X \cdot I_{p-2} - l_x) = X^{p-1} - (\mathrm{tr}(x)) \cdot X^{p-2} + \cdots + (-1)^{p-1}\det(l_x).$$

We get $\mathrm{tr}(x) = -a_{p-2}$ and $\mathrm{N}(x) = (-1)^{p-1}a_0$. Since $K = \mathbb{Q}(x)$, $F(X) = (X - x_1)(X - x_2)\cdots(X - x_{p-1})$ by the definition of the x_i. Therefore, $\mathrm{tr}(x) = -a_{p-2} = x_1 + x_2 + \cdots x_{p-1}$, and $\mathrm{N}(x) = (-1)^{p-1}a_0 = x_1 x_2 \cdots x_{p-1}$ as desired.

Now we do the general case. Let m be the dimension of K over $\mathbb{Q}(x)$ and suppose $m \neq 1$. It suffices to show that the roots of P, the characteristic polynomial, will be those of F repeated m times. Let $y_1, \ldots y_r \in \mathbb{Q}(x)$ be a basis for $\mathbb{Q}(x)$ over $\mathbb{Q}$, and $z_1, \ldots, z_m \in K$ be a basis for K over $\mathbb{Q}(x)$.

Exercise 2.3: Show $y_1 z_1, y_1 z_2, \ldots, y_1 z_m, y_2 z_1, \ldots, y_2 z_m, \ldots, y_r z_1, \ldots, y_r z_m$ is a basis for K over $\mathbb{Q}$, so $rm = p - 1$.

Let $M = (a_{ik})$ be the matrix for l_x in $\mathbb{Q}(x)$ w.r.t. the y_i, i.e. $xy_i = \sum_{k=1}^{r} a_{ik} y_k$. Then

$$x(y_i z_j) = \left(\sum_{k=1}^{r} a_{ik} y_k\right) z_j = \sum_{k=1}^{r} a_{ik}(y_k z_j).$$

Thus the matrix M_1 for l_x in K (considered as a vector space over $\mathbb{Q}$) w.r.t. to the basis $\{y_k z_j\}$, $k = 1, \ldots, r$ and $j = 1, \ldots, m$, looks like

$$M_1 = \begin{pmatrix} M & 0 & \cdots & 0 \\ 0 & M & \cdots & 0 \\ \vdots & \vdots & & \vdots \\ 0 & 0 & \cdots & M \end{pmatrix},$$

i.e. M occurs m times as an $r \times r$ block along the diagonal in M_1. We see that the matrix $X \cdot I_{p-1} - M_1$ consists of m blocks along the diagonal, each of size $r \times r$, and each looks like $X \cdot I_r - M$. Hence

$$\det(X \cdot I_{p-1} - M_1) = [\det(X \cdot I_r - M)]^m. \tag{5}$$

The left hand side of (5) is $P(X)$, and by the first part of the proof, $\det(X \cdot I_r - M) = F(X)$, the minimal polynomial. □

Now we can compute some more traces in $K = \mathbb{Q}(\varsigma)$. Since the cyclotomic polynomial, $F(X) = X^{p-1} + \cdots + X + 1$, is irreducible, it is the minimal polynomial of ς in K. Since $\mathrm{tr}(\varsigma) = -a_{p-2}$, we get $\mathrm{tr}(\varsigma) = -1$. In addition, by Exercise 2.2, part iii), $\mathrm{tr}(1) = p - 1$. Now $F(X)$ is also the minimal polynomial for ς^i, so $\mathrm{tr}(\varsigma^i) = -1$ for $i = 1, 2, \ldots, p-1$. Hence

$$\mathrm{tr}(1 - \varsigma) = \mathrm{tr}(1 - \varsigma^2) = \cdots = \mathrm{tr}(1 - \varsigma^{p-1}) = p. \tag{6}$$

Exercise 2.4: Show $\mathrm{N}(1-\varsigma) = p$. (**Hint:** Look at the proof of Proposition 2.2.)

Now by (4), $\mathrm{N}(1-\varsigma)$ is also the product of the roots of the minimal polynomial of $1-\varsigma$.

Exercise 2.5: Show that the roots of the minimal polynomial of $1-\varsigma$ are $1-\varsigma, 1-\varsigma^2, \ldots, 1-\varsigma^{p-1}$. (**Hint:** Same as the previous exercise.)

Therefore

$$p = (1-\varsigma)(1-\varsigma^2)\cdots(1-\varsigma^{p-1}). \tag{7}$$

Notice that since $(1-\varsigma^i)$ is in $R(=\mathbb{Z}[\varsigma])$, (7) shows that p, which was a prime in $\mathbb{Z}$, is far from a prime in R. (We know that the factors are not units. How?) We must now examine more closely how R sits inside K. It is time to make some more definitions.

Definition 2.4: A polynomial P in $\mathbb{Z}[X]$ is said to be *monic* if its leading coefficient is 1, i.e. $P(X) = X^n + \cdots + a_0$. An element $x \in K$ is said to be *integral* or an *integer* if it is the root of a monic polynomial in $\mathbb{Z}[X]$.

Proposition 2.3: *The set A of integral elements of K forms a subring of K.*

The proof will use the following:

Lemma 2.1: *$x \in K$ is integral iff $\mathbb{Z}[x]$ is a finitely generated abelian group.*

Proof: Suppose x is integral and is a root of

$$P(X) = X^n + a_{n-1}X^{n-1} + \cdots + a_1 X + a_0$$

with the $a_i \in \mathbb{Z}$. Suppose that M is the $\mathbb{Z}$-submodule of K generated by $\{1, x, \ldots x^{n-1}\}$. Since $P(x) = 0$, $x^n \in M$. We see by induction that $x^i \in M$ for $i \geq 0$. But $\mathbb{Z}[x]$ is the $\mathbb{Z}$-module generated by the powers of x. Hence $\mathbb{Z}[x] = M$ and therefore is finitely generated.

Conversely suppose $\mathbb{Z}[x]$ is finitely generated. Let $y_1, y_2, \ldots, y_n \in \mathbb{Z}[x]$ be a set of generators. $x \in \mathbb{Z}[x] \Rightarrow xy_i \in \mathbb{Z}[x]$ for $i = 1, 2, \ldots, n$. Hence

$$xy_i = \sum_{j=1}^{n} a_{ij} y_j \tag{8}$$

for $i = 1, 2, \ldots, n$ and $a_{ij} \in \mathbb{Z}$ for $1 \leq i, j \leq n$. Another way to write (8) is

$$\sum_{j=1}^{n} (\delta_{ij} x - a_{ij}) y_j = 0 \quad \text{for } i = 1, 2, \ldots, n. \tag{9}$$

Think of (9) as a homogenous set of n linear equations in the y_i's. Since the hypothesis is that we can find the y_i's, we must be able to solve the system (9), which means that

$$\det(\delta_{ij} x - a_{ij}) = 0.$$

But $\det(\delta_{ij} X - a_{ij})$ is a polynomial in $\mathbb{Z}[X]$, and a little reflection will convince you it is monic. □

Remark: The second half of the proof actually shows that any element of $\mathbb{Z}[x]$ is integral.

Now we give the proof of Proposition 2.3.

Proof: We need to show that if x and y are integral, then so are $x \pm y$ and xy. Now $x \pm y$ and xy are in $\mathbb{Z}[x, y]$, and x integral implies that $\mathbb{Z}[x]$ is finitely generated over ς. We can write $\mathbb{Z}[x, y] = (\mathbb{Z}[x])\,[y]$.

Exercise 2.6: Show that $\mathbb{Z}[x, y]$ is finitely generated as a $\mathbb{Z}$-module.

By the remark after Lemma 2.1 we are done. □

The next proposition tells us about the norms and traces of integeral elements.

Proposition 2.4: *Let $x \in K$ be integral. Then the coefficients of its characteristic polynomial P are in $\mathbb{Z}$. In particular,* $\mathrm{tr}(x)$ *and* $\mathrm{N}(x) \in \mathbb{Z}$.

Proof: Now the coefficients of P are certainly in $\mathbb{Q}$. If we can show they are integral, they will be in $\mathbb{Z}$. Since the coefficients of P are the sum of products of the roots, if we show the roots of P are integral, then the coefficients will be integral by Proposition 2.3. Let $x_1, \ldots, x_n$ be the roots of P. By Proposition 2.2, $x_1, \ldots, x_n$ are just the roots of the minimal polynomial of x repeated a number of times.

Exercise 2.7: Let f be an irreducible polynomial in $\mathbb{Q}[X]$. Show that if one root of f is integral, then all the roots of f are integral.

The exercise finishes the proof. □

Now we have $K = \mathbb{Q}(\varsigma)$ and two subrings, $R(= \mathbb{Z}[\varsigma])$ and A, the subring of integral elements. Notice that $R \subset A$ since R is certainly a finitely generated $\mathbb{Z}$-module, or, if you prefer, notice that $1, \varsigma, \varsigma^2, \ldots, \varsigma^{p-1} \in A$ since they are the roots of $X^p - 1$. In general, if you adjoin an element $x \in \mathbb{C}$ to $\mathbb{Q}$ and look at the subring of integral elements of that field, it may not be $\mathbb{Z}[x]$. For example, on page 35 of [S], you can see this is not the case for $x = \sqrt{5}$. Fortunately in this case it is, i.e. $A = \mathbb{Z}[\varsigma]$. To prove this, we first need a lemma.

Lemma 2.2: *i)* $(1-\varsigma)A \cap \mathbb{Z} = p \cdot \mathbb{Z}$.

ii) For any $y \in A$, $\operatorname{tr}[(1-\varsigma)y] \in p \cdot \mathbb{Z}$.

Proof: Equation (7) says $p \in (1-\varsigma)A$. Since $(1-\varsigma)A$ is an ideal in A, $(1-\varsigma)A \cap \mathbb{Z}$ is an ideal in $\mathbb{Z}$ which must contain the ideal $p\mathbb{Z}$. Now $p\mathbb{Z}$ is a maximal ideal in $\mathbb{Z}$ since p is prime. If i) does not hold, we must have $(1-\varsigma)A \cap \mathbb{Z} = \mathbb{Z}$. If this were the case, then $(1-\varsigma)$ would have to have an inverse in A. But by Exercise 2.4, $\mathrm{N}(1-\varsigma) = \det(l_{1-\varsigma}) = p$. Therefore if $(1-\varsigma)$ had an inverse in A, p would have to have an inverse in $\mathbb{Z}$ which is absurd.

We use i) to prove ii).

Exercise 2.8: Show that the roots of the minimal polynomial of $y(1-\varsigma)$ are $y_j(1-\varsigma^j)$ for some $y_j \in A$.

Since $1-\varsigma^j = (1-\varsigma)(1+\varsigma+\varsigma^2+\cdots+\varsigma^{j-1})$, each root of the minimal polynomial of $y(1-\varsigma)$ is itself a multiple of $(1-\varsigma)$. By equation (3), we see that $\operatorname{tr}((1-\varsigma)y) \in (1-\varsigma)A$. But for $y \in A, (1-\varsigma)y$ is integral by Proposition 2.3. By Proposition 2.4, $\operatorname{tr}((1-\varsigma)y) \in \mathbb{Z}$, hence by i)

$$\operatorname{tr}((1-\varsigma)y) \in (1-\varsigma)A \cap \mathbb{Z} = p \cdot \mathbb{Z},$$

and we are done. □

Theorem 2.2 : *The ring of integral elements in* $K = \mathbb{Q}(\varsigma)$ *is* $R = \mathbb{Z}[\varsigma]$, *i.e.* $A = R$.

Proof: We know that $R \subset A$. Now let $x \in A$. Since $1, \varsigma, \varsigma^2, \ldots, \varsigma^{p-2}$ is a basis for K over $\mathbb{Q}$, we can write

$$x = a_0 + a_1\varsigma + a_2\varsigma^2 + \cdots + a_{p-2}\varsigma^{p-2} \tag{10}$$

with $a_i \in \mathbb{Q}$ for $i = 0, 1, 2, \ldots, p-2$. Therefore

$$(1-\varsigma)x = a_0(1-\varsigma) + a_1(\varsigma - \varsigma^2) + \cdots + a_{p-2}(\varsigma^{p-2} - \varsigma^{p-1}). \tag{11}$$

Recall that $\mathrm{tr}(\varsigma) = -1$ and $\mathrm{tr}(1) = p$, so $\mathrm{tr}(\varsigma^j - \varsigma^{j+1}) = 0$ for $j = 1, 2, \ldots, p-2$. Hence (11) says

$$\mathrm{tr}\left[(1-\varsigma)x\right] = \mathrm{tr}\left[a_0(1-\varsigma)\right] = a_0 p.$$

But by part ii) of Lemma 2.2, $\mathrm{tr}\left[(1-\varsigma)x\right] \in p\mathbb{Z}$, so $a_0 p \in p\mathbb{Z}$ and hence $a_0 \in \mathbb{Z}$.

Now $\varsigma^{-1} = \varsigma^{p-1} \in A$. Therefore

$$(x - a_0)\varsigma^{-1} = a_1 + a_2\varsigma + \cdots + a_{p-2}\varsigma^{p-3} \in A. \tag{12}$$

Multiply (12) by $(1-\varsigma)$ and take traces. We get $a_1 \in \mathbb{Z}$. Continuing the argument, we get $a_0, a_1, a_2, \ldots, a_{p-2} \in \mathbb{Z}$, and so by (11), $x \in \mathbb{Z}[\varsigma]$. □

Before moving on to the next section which discusses modules over R, we would like to offer a brief presentation of how ideals in R factor into primes as mentioned above. As usual, we have some lemmas first.

Lemma 2.3: *Every ideal in R contains a product of prime ideals.*

Proof: Let S be the set of ideals for which this statement is false. Since R is finitely generated over $\mathbb{Z}$, any ideal of R is finitely generated over $\mathbb{Z}$. Therefore if $S \neq \emptyset$, S must have a maximal element, say $\mathfrak{a}$. Obviously $\mathfrak{a}$ is a proper non-prime ideal. Furthermore if $\mathfrak{a} \subset \mathfrak{b}$, then $\mathfrak{b}$ contains a product of primes.

Since $\mathfrak{a}$ is non-prime, there are $a, b \in R$ s.t. $ab \in \mathfrak{a}$, but neither a nor b in $\mathfrak{a}$. Now both $\mathfrak{a} \subset \mathfrak{a} + aR$ and $\mathfrak{a} \subset \mathfrak{a} + bR$, so both $\mathfrak{a} + aR$ and $\mathfrak{a} + bR$ contain products of primes. But

$$(\mathfrak{a} + aR)(\mathfrak{a} + bR) \subset \mathfrak{a} \cdot \mathfrak{a} + a\mathfrak{a} + b\mathfrak{a} + abR \subset \mathfrak{a},$$

so $\mathfrak{a}$ also contains a product of primes which is a contradiction. Hence $S = \emptyset$. □

Lemma 2.4: *Let $\mathfrak{a}$ be a proper ideal of R and set $\mathfrak{a}^{-1} = \{x \in K : x\mathfrak{a} \subset R\}$. Then R is a proper subset of $\mathfrak{a}^{-1}$.*

Proof: Clearly $R \subset \mathfrak{a}^{-1}$, so it only remains to show $R \neq \mathfrak{a}^{-1}$. Pick $a \in \mathfrak{a} - \{0\}$. By Lemma 2.3, we can find prime ideals $\mathfrak{p}_1, \ldots, \mathfrak{p}_n$ of R s.t.

$$\mathfrak{p}_1\mathfrak{p}_2 \cdots \mathfrak{p}_n \subset aR \subset \mathfrak{a},$$

Pick $\mathfrak{p}_1, \ldots, \mathfrak{p}_n$ s.t. n is minimal.

Exercise 2.9: Show that any proper ideal of R is contained in a maximal ideal of R.

Let $\mathfrak{a} \subset \mathfrak{m}$, $\mathfrak{m}$ a maximal ideal. Hence

$$\mathfrak{p}_1\mathfrak{p}_2\cdots\mathfrak{p}_n \subset \mathfrak{m}.$$

Since $\mathfrak{m}$ is also prime, $\mathfrak{m} = \mathfrak{p}_i$ for some i, say $\mathfrak{m} = \mathfrak{p}_1$. We get

$$\mathfrak{m}\mathfrak{p}_2\cdots\mathfrak{p}_n \subset aR \subset \mathfrak{a} \subset \mathfrak{m} \subset R. \tag{13}$$

Since n is minimal, $\mathfrak{p}_2\cdots\mathfrak{p}_n \not\subset aR$, so we can find $b \in \mathfrak{p}_2\cdots\mathfrak{p}_n$ s.t. $b \notin aR$. Let $\lambda = b/a$. Clearly $\lambda \in K$ and $\lambda \notin R$, but

$$\lambda\mathfrak{a} = (b/a)\mathfrak{a} \subset (b/a)\mathfrak{m} = (1/a)\mathfrak{m}b \subset (1/a)\mathfrak{m}\mathfrak{p}_2\cdots\mathfrak{p}_n,$$

since $b \in \mathfrak{p}_2\cdots\mathfrak{p}_n$. Hence by (13) $\lambda\mathfrak{a} \subset R$, and $\lambda \in \mathfrak{a}^{-1}$ by definition. □

Theorem 2.3 (Fundamental Theorem of Arithmetic for R): *Every proper ideal $\mathfrak{a}$ of R is a product of prime ideals. Furthermore this decomposition is unique up to rearrangement.*

Proof: By Lemma 2.3, $\mathfrak{a}$ contains a product of primes, $\mathfrak{p}_1, \ldots, \mathfrak{p}_n$. Again pick a maximal ideal $\mathfrak{m}$ with $\mathfrak{a} \subset \mathfrak{m}$. As in the proof of Lemma 2.4 we can assume that $\mathfrak{m} = \mathfrak{p}_1$, so

$$\mathfrak{m}^{-1} \cdot (\mathfrak{m}\mathfrak{p}_2\cdots\mathfrak{p}_n) = \mathfrak{p}_2\cdots\mathfrak{p}_n \subset \mathfrak{m}^{-1}R \subset R.$$

Now suppose by induction that we have shown that any ideal which contains a product of fewer than n prime ideals is itself the product of prime ideals. Then we have

$$\mathfrak{m}^{-1}R = \mathfrak{p}_1\mathfrak{p}_2\cdots\mathfrak{p}_r$$

with gp_i prime. Therefore

$$\mathfrak{a} = \mathfrak{m}\mathfrak{m}^{-1}R = \mathfrak{m}\mathfrak{p}_1\mathfrak{p}_2\cdots\mathfrak{p}_r$$

as desired.

Exercise 2.10: Prove the uniqueness part of the theorem. (**Hint:** It is easy. Look again at the proof of Lemma 2.4).

The exercise concludes the proof. □

We will need some consequences of this theorem in the sequel. The first is a trivial consequence of the theorem.

Corollary 2.1: *Let $\mathfrak{a}$ and $\mathfrak{b}$ be ideals. Then $\mathfrak{a} \subset \mathfrak{b}$ iff there is an ideal $\mathfrak{c}$ s.t. $\mathfrak{a} = \mathfrak{b}\mathfrak{c}$.*

Next we use the theorem to prove a lemma which is needed for the famous "Chinese Remainder Theorem."

Lemma 2.5: *Let $\mathfrak{a}$ and $\mathfrak{b}$ be ideals, and suppose that*

$$\mathfrak{a} = \mathfrak{p}_1^{a_1} \cdots \mathfrak{p}_m^{a_m} \text{ and } \mathfrak{b} = \mathfrak{p}_1^{b_1} \cdots \mathfrak{p}_m^{b_m}$$

where the $\mathfrak{p}_i$ are distinct primes, and $a_i, b_i \in \mathbb{Z}$ and $a_i, b_i \geq 0$. Then

$$\mathfrak{a} + \mathfrak{b} = \prod_{i=1}^{n} \mathfrak{p}_i^{\min(a_i, b_i)} \tag{14}$$

and

$$\mathfrak{a} \cap \mathfrak{b} = \prod_{i=1}^{n} \mathfrak{p}_i^{\max(a_i, b_i)} \tag{15}$$

Exercise 2.11: Prove this lemma, and then show that $\mathfrak{a}\mathfrak{b} = (\mathfrak{a}+\mathfrak{b})(\mathfrak{a}\cap\mathfrak{b})$.

Definition 2.5: We say that two ideals $\mathfrak{a}$ and $\mathfrak{b}$ are *relatively prime* if $\mathfrak{a} + \mathfrak{b} = R$.

Exercise 2.12: i) Show that $\mathfrak{a}$ and $\mathfrak{b}$ are relatively prime iff they have no common factor.

ii) Show that if $\mathfrak{a}$ and $\mathfrak{b}$ are relatively prime, then $\mathfrak{a}\mathfrak{b} = \mathfrak{a} \cap \mathfrak{b}$.

iii) Show that $\mathfrak{a}_1, \ldots, \mathfrak{a}_n$ are pairwise relatively prime iff

$$\mathfrak{a}_i + \prod_{i \neq j} \mathfrak{a}_j = R$$

for $i = 1, 2, \ldots, n$.

Theorem 2.4 (Chinese Remainder Theorem): *Let $\mathfrak{a}_1, \ldots, \mathfrak{a}_n$ be a collection of pairwise relatively prime ideals, and let $a_1, \ldots, a_n$ be arbitrary elements of R. Then there is an element $a \in R$ s.t.*

$$a \equiv a_i \ (\mathrm{mod}\ \mathfrak{p}_i) \qquad \text{for } i = 1, 2, \ldots n.$$

In other words, a and a_i determine the same element in $R/\mathfrak{a}_i$. Furthermore a is unique modulo $(\mathfrak{a}_1 \cdots \mathfrak{a}_n)$.

Proof: By Exercise 2.12, part iii), we can find elements $b_i \in \mathfrak{a}_i$ and $b_i' \in \prod_{j \neq i} \mathfrak{a}_i$ s.t. $b_i + b_i' = 1$ for $i = 1, 2, \ldots, n$. Let

$$a = b_1' a_1 + \cdots + b_n' a_n.$$

Hence $a \equiv b_i' a_i \pmod{\mathfrak{a}_i}$. Since $b_i + b_i' = 1$, we must have $b_i' \equiv 1$ (mod $\mathfrak{a}_i$), and therefore $a \equiv a_i \pmod{\mathfrak{a}_i}$. If $a' \equiv a_i \pmod{\mathfrak{a}_i}$, then $a' - a \equiv 0 \pmod{\mathfrak{a}_i}$, so $a' - a \in \mathfrak{a}_1 \cap \cdots \cap \mathfrak{a}_n$ which is $\mathfrak{a}_1 \cdots \mathfrak{a}_n$ by part ii) of Exercise 2.12. □

Theorem 2.5 (Hecke): *Let $\mathfrak{a}$ and $\mathfrak{b}$ be ideals. Then there is an ideal $\mathfrak{c}$ relatively prime to $\mathfrak{b}$ s.t. $\mathfrak{a}\mathfrak{c}$ is principal.*

Proof: Write

$$\mathfrak{a} = \mathfrak{p}_1^{a_1} \cdots \mathfrak{p}_m^{a_m} \text{ and } \mathfrak{b} = \mathfrak{p}_1^{b_1} \cdots \mathfrak{p}_m^{b_m}$$

where the $\mathfrak{p}_i$ are distinct primes, and $a_i, b_i \in \mathbb{Z}$ and $a_i, b_i \geq 0$. Pick $x_i \in \mathfrak{p}_i^{a_i}$ s.t. $x_i \notin \mathfrak{p}_i^{a_i+1}$. By the Chinese Remainder theorem, we can find $x \in R$ s.t.

$$x \equiv x_i \pmod{\mathfrak{p}_i^{a_i+1}} \tag{16}$$

for $i = 1, 2, \ldots, n$. Then

$$x \equiv 0 \pmod{\mathfrak{p}_i^{a_i}},$$

so $x \in \mathfrak{p}_i^{a_i}$, but $x \notin \mathfrak{p}_i^{a_i+1}$. Hence $x \in \mathfrak{p}_1^{a_1} \cap \cdots \cap \mathfrak{p}_m^{a_m}$ which by part ii) of Exercise 2.12 is $\mathfrak{p}_1^{a_1} \cdots \mathfrak{p}_m^{a_m}$, so $x \in \mathfrak{a}$.

Now equation (16) says that in the prime decomposition of xR, the prime $\mathfrak{p}_i$ appears precisely to the power a_i, i.e. $xR \subset \mathfrak{a}$, so $xR = \mathfrak{a}\mathfrak{c}$ for some ideal $\mathfrak{c}$, and each $\mathfrak{p}_i$ appears to the 0th power in the prime decomposition of $\mathfrak{c}$. Now we show that equation (14) implies

$$xR + \mathfrak{a}\mathfrak{b} = \mathfrak{a}. \tag{17}$$

To see this note that $\mathfrak{a}\mathfrak{b} = \prod \mathfrak{p}_i^{a_i+b_i}$ and $a_i \leq a_i + b_i$, so no prime that appears in the prime decomposition of $\mathfrak{c}$ can contribute, since it will appear to the 0th power in $\mathfrak{a}\mathfrak{b}$. Substituting $\mathfrak{a}\mathfrak{c}$ for xR in (17) yields

$$\mathfrak{a}\mathfrak{c} + \mathfrak{a}\mathfrak{b} = \mathfrak{a}.$$

Exercise 2.13: Show this implies $\mathfrak{c} + \mathfrak{b} = R$.

The proof of the theorem concludes by observing that $\mathfrak{a}\mathfrak{c}$ is xR, which is certainly principal. □

As an application of this theorem we show that ideals that are not principal (generated by one element) are generated (as an ideal) by two elements, so they are not so far from principal.

Corollary 2.2: *Let $\mathfrak{a}$ be any ideal of R. Then there are elements a and b in R s.t. $\mathfrak{a} = aR + bR$.*

Proof: Let a be an arbitrary non-zero element of $\mathfrak{a}$. By the theorem we can pick an ideal $\mathfrak{b}$ with $\mathfrak{b} + aR = R$ and with $\mathfrak{a}\mathfrak{b}$ a principal ideal, say $\mathfrak{a}\mathfrak{b} = bR$. Then

$$aR + bR = AR + \mathfrak{a}\mathfrak{b} \subset \mathfrak{a}$$

since both aR and $\mathfrak{a}\mathfrak{b}$ are subsets of $\mathfrak{a}$.

On the other hand, since $\mathfrak{b} + aR = R$, we can write $1 = x + ay$ for $x \in \mathfrak{b}$ and $y \in R$. Hence if z is arbitrary in $\mathfrak{a}$, we have

$$z = zx + zay \in \mathfrak{a}\mathfrak{b} + aR = bR + aR.$$

Therefore $\mathfrak{a} \subset aR + bR$. □

Another application of this theorem will appear at the end of the next section (Lemma 3.4).

3. Modules over the Cyclotomic Ring

The next step in finding the faithful integral representations of $\Phi(= \mathbb{Z}_p)$ or what is the same thing, finding the faithful Φ-modules, is to look at modules over $R(= \mathbb{Z}[\varsigma])$. Some examples of R-modules are K and any ideal $\mathfrak{a}$ of R. All R-modules are Φ-modules (see Exercise 4.2).

Recall (Definition 2.1) that two ideals, $\mathfrak{a}$ and $\mathfrak{b}$, are equivalent or in the same ideal class if we can find $x, y \in R$ with $x\mathfrak{a} = y\mathfrak{b}$. Another way of saying this is to say we can find $\alpha \in K$ with $\alpha \cdot \mathfrak{a} = \mathfrak{b}$, i.e. $\alpha = x/y$.

Suppose M is any R-module. Then we can form the abelian group $K \otimes_R M$ by regarding K as a right R-module and M, as usual, as a left R-module. We can make $K \otimes_R M$ into a vector space over K by setting

$$k' \cdot (k \otimes m) = (k'k) \otimes m \tag{18}$$

for $k, k' \in K$ and $m \in M$. The dimension of $K \otimes_R M$ is defined to be the *rank* of M over R.

Exercise 3.1: Show that the R-module M can be identified with the R-submodule $1 \otimes_R M$ of $K \otimes_R M$.

Proposition 3.1: *Let $\mathfrak{a}$ and $\mathfrak{b}$ be ideals of R. Then they are isomorphic as R-modules iff they are in the same ideal class.*

Proof: If $\mathfrak{a}$ and $\mathfrak{b}$ are in the same ideal class, there is $\alpha \in K$ s.t.

$$\alpha\mathfrak{a} = \mathfrak{b}. \tag{19}$$

We can define a map $\varphi : \mathfrak{a} \to \mathfrak{b}$ by $\varphi(x) = \alpha x$ for $x \in \mathfrak{a}$. Ordinarily, αx would not even be in R, but (19) tells us that it is not only in R but even in $\mathfrak{b}$. It is easy to check that φ is an R-isomorphism.

Conversely, let $\varphi : \mathfrak{a} \to \mathfrak{b}$ be an R-isomorphism. Then we can extend φ to a vector space isomorphism

$$\varphi' : K \otimes_R \mathfrak{a} \to K \otimes_R \mathfrak{b}$$

by

$$\varphi'(k \otimes x) = k \otimes \varphi(x) \tag{20}$$

for $k \in K$ and $x \in \mathfrak{a}$.

Exercise 3.2: Let a_0 and a_1 be arbitrary in $\mathfrak{a}$. Show that $a_0^{-1} \otimes a_0 = a_1^{-1} \otimes a_1$ in $K \otimes_R \mathfrak{a}$. (**Hint:** Show $a_1(a_0^{-1} \otimes a_0) = a_1(a_1^{-1} \otimes a_1) = 1 \otimes a_1$. Be careful, 1 is probably not in $\mathfrak{a}$.)

Now fix $a \in \mathfrak{a}$. We have by (20) that

$$\varphi'(a^{-1} \otimes a) = a^{-1} \otimes \varphi(a). \tag{21}$$

Since $a^{-1} \otimes a$ is independent of a, $a^{-1} \otimes \varphi(a)$ is also independent of a.

Claim: For any $x \in \mathfrak{a}$,

$$\varphi(x) = \frac{\varphi(a)}{a} \cdot x$$

or, equivalently, $a\varphi(x) = \varphi(a)x$.

We identify $a\varphi(x)$ with $1 \otimes a\varphi(x)$ and $\varphi(a)x$ with $1 \otimes \varphi(a)x$. We get

$$\begin{aligned} 1 \otimes a\varphi(x) &= a \otimes \varphi(x) && \text{since } a \in \mathfrak{a} \subset R \\ &= a(1 \otimes \varphi(x)) && \text{by (18)}, \end{aligned}$$

and

$$
\begin{aligned}
1 \otimes \varphi(a)x &= x \otimes \varphi(a) && \text{since } x \in \mathfrak{a} \subset R \\
&= xa(a^{-1} \otimes \varphi(a)) && \text{by (18)} \\
&= xa\varphi'(a^{-1} \otimes a) && \text{by (20)} \\
&= xa\varphi'(x^{-1} \otimes x) && \text{by Exercise 9.5} \\
&= xa(x^{-1} \otimes \varphi(x)) && \text{by (21)} \\
&= a(1 \otimes \varphi(a)) && \text{by (18).}
\end{aligned}
$$

The claim thus follows, and we see that φ is nothing more than multiplication by the element of K given by $\alpha = \varphi(a)/a$, so $\mathfrak{a}$ and $\mathfrak{b}$ are in the same ideal class. □

We are aiming at a theorem which says that "reasonable" modules over R are merely direct sums of ideals. The next few lemmas will start us in the right direction and also tell us what "reasonable" should be.

Lemma 3.1: *Let M be a finitely generated R-submodule of the R-module K. Then M is isomorphic to an ideal of R.*

Proof: Let $m_1, \ldots, m_k$ be a set of generators for M. Since $M \subset K$, we can write $m_i = a_i/b_i$ with $a_i, b_i \in R$, for $i = 1, \ldots, k$. Let $b = \prod_{i=1}^{k} b_i$. Then $bM \subset R$, bM is isomorphic to M, and bM is an ideal of R. □

Remarks: i) We call the process of replacing a finitely generated R-submodule of K by an ideal of R, *clearing denominators.* Notice that it is crucial to start with a finitely generated R-module, so one part of "reasonable" should be "finitely generated."

ii) The finitely generated R-submodules of K are the "fancier ideals" or *fractional ideals* mentioned after Definition 2.1. If we use them, we would get a somewhat simpler treatment of R-modules (see [28], Section 22). Because we are trying to keep the introduction of new ideas to a minimum, we omit them.

Definition 3.1: We say an R-module M is *torsionfree* if whenever $am = 0$ for $a \in R$ and $m \in M$ we have either $a = 0$ or $m = 0$.

"Torsionfree" is the other part of "reasonable." The proof that "reasonable" modules are direct sums of ideals will be by induction on the rank. The next lemma begins the induction.

Lemma 3.2: *Let M be a torsionfree finitely generated R-module of rank 1. Then M is isomorphic to an ideal of R.*

Proof: By the definition of rank, $K \otimes_R M$ is a vector space over K of dimension 1. Recall that we can identify M with $1 \otimes_R M \subset K \otimes_R M$. Take $m \in M - \{0\}$ so that $K \otimes_R M$ is generated by $1 \otimes m$. Let

$$I = \{\alpha \in K : \alpha \otimes m \in 1 \otimes M\}.$$

Then I is an R-submodule of K, and the map $\alpha \mapsto \alpha \otimes m$ is an R-isomorphism from I to $1 \otimes M$ since M is torsionfree. Since M is finitely generated, so is I. By Lemma 3.1, M and I are isomorphic to some ideal of R. □

Lemma 3.3: *Let M be a finitely generated torsionfree R-module. Let $\alpha \in K$, and suppose $\alpha \otimes_R M \subset 1 \otimes_R M$. Then $\alpha \in R$.*

Proof: Pick $m \in M - \{0\}$ and let

$$I = \{\beta \in K : \beta \otimes m \in 1 \otimes M\}.$$

We copy the proof of the previous lemma to conclude that I is a finitely generated R-submodule of K. Notice that this time I is not isomorphic to M, but it is isomorphic to the module of rank 1 generated by m. In any case, we clearly have $R \subset I$.

Let $\alpha \in K$ with $\alpha \otimes M \subset 1 \otimes M$. Since $R[\alpha]$ is the smallest ring containing R and α, and clearly $\alpha \in I$, we see that $R[\alpha] \subset I$. Again using the proof of Lemma 3.1, we see that I is isomorphic to an ideal of R, so $R[\alpha]$ is isomorphic to an ideal $\mathfrak{a}$ of R.

Exercise 3.3: Show $\mathbb{Z}[\alpha]$ is a finitely generated abelian group. (**Hint:** As abelian groups, just about everything in sight is finitely generated, e.g. $R, \mathfrak{a}, R[\alpha]$, etc.)

Hence by Lemma 2.1, α is an integral element, and so by Theorem 2.2, $\alpha \in R (= \mathbb{Z}[\varsigma])$. □

We need one more notion before beginning the proof of the theorem we are aiming towards.

Definition 3.2: Let $\mathcal{R}$ be any ring (commutative, of course) and M any $\mathcal{R}$-module. A submodule N of M is said to be *pure* if for each $x \in \mathcal{R}$, $xN = N \cap xM$.

Exercise 3.4: i) Show N is pure iff for $x \in \mathcal{R}$ and $m \in M$, $xm \in N \Rightarrow xm = xn$ for some $n \in N$.

ii) If n is a direct summand of M, show that N is pure.

iii) Suppose M is torsionfree. Show that N is pure iff for all $m \in M$ and $x \in \mathcal{R} - \{0\}$ if $xm \in N$ then $m \in N$.

iv) Show that if M/N is torsionfree, then N is pure. Show also that if M is torsionfree, and N is a pure submodule, then M/N is torsionfree.

For the next two parts of this exercise we need some hypothesis on the ring $\mathcal{R}$. For example, you can assume $\mathcal{R}$ is a P.I.D. or Dedekind if you know what these mean. For our purposes it suffices to restrict to the case $\mathcal{R} = R = \mathbb{Z}[\varsigma]$.

v) Let V_0 be a finite dimensional vector space over K with basis $v_1, \ldots, v_n$. Let V be the R-module

$$V = Rv_1 \oplus \cdots \oplus Rv_n.$$

Suppose W_0 is a subspace of V_0. Show that the module $W_0 \cap V$ is a pure submodule of V.

vi) Let S be an arbitrary subset of V (notation as above). Let W_0 be the subspace of V_0 generated by S. Show that $W_0 \cap V$ is the unique minimal pure submodule of V containing S.

Keeping the same notation, we say that $W_0 \cap V$ is the *pure submodule of V generated by S*. If M is any R-module, we can take for V the vector space $K \otimes_R M$. Then $1 \otimes S$ will generate a subspace W_0 of $K \otimes_R M$, and $N = W_0 \cap (1 \otimes M)$ can be identified with a pure submodule of M. We say N is the *pure submodule of M generated by S*.

Now on to our theorem. If $\mathfrak{a}_1, \ldots, \mathfrak{a}_n$ are ideals in R, then $\mathfrak{a}_1 \oplus \cdots \oplus \mathfrak{a}_n$ is a finitely generated torsionfree R-module of rank n. We now show the converse.

Theorem 3.1 : *Let M be a finitely generated torsionfree R-module of rank n. Then there are n ideals $\mathfrak{a}_1, \ldots, \mathfrak{a}_n$ of R s.t.*

$$M \approx \mathfrak{a}_1 \oplus \cdots \oplus \mathfrak{a}_n.$$

Proof: As we have remarked, the proof is by induction on n. Lemma 3.2 is the result for $n = 1$, and we assume the theorem is true for modules of rank less than n. As usual, we identified M with $1 \otimes_R M$ in $K \otimes_R M$. Fix $m \in M - \{0\}$ and let W_0 be the subspace of $K \otimes_R M$ generated by $1 \otimes m$.

Then as in Exercise 3.4, part vi), $W_0 \cap (1 \otimes_R M)$ can be identified with a pure submodule N of M ("the pure submodule of M generated by m"). By part iv) of Exercise 3.4, M/N is torsionfree over R, finitely generated, and of rank $n-1$. By induction, we get ideals $\mathfrak{a}_1, \ldots, \mathfrak{a}_{n-1}$ of R s.t.

$$M/N \approx \mathfrak{a}_1 \oplus \cdots \oplus \mathfrak{a}_{n-1}. \tag{22}$$

Now suppose we can show that N is a direct summand (over R) of M, i.e. $M = N \oplus T$ for some R-module T. Then by (22)

$$T \approx \mathfrak{a}_1 \oplus \cdots \oplus \mathfrak{a}_{n-1},$$

and since N is a torsionfree module of rank 1, by Lemma 3.2, N is isomorphic to an ideal of R, so we will be done.

Equation (22) implies that there is a homomorphism $\varphi : M \to \mathfrak{a}_1 \oplus \cdots \oplus \mathfrak{a}_{n-1}$ with $\ker\varphi = N$. Put $M_i = \varphi^{-1}(\mathfrak{a}_i)$ for $i = 1, \ldots, n-1$. Let $\varphi_i = \varphi | M_i$, so $\varphi_i : M_i \to \mathfrak{a}_i$ and $\ker\varphi_i = N$ for $i = 1, \ldots, n-1$. We want to find a pure submodule T_i of M_i with $M_i = T_i \oplus N$.

Let $\mathfrak{a}_i^{-1} = \{x \in K : x\mathfrak{a}_i \subset R\}$ for $i = 1, \ldots, n-1$. Fix i for a while.

Exercise 3.5: Show there are elements $a_1, \ldots, a_r \in \mathfrak{a}_i^{-1}$ and $b_1, \ldots, b_r \in \mathfrak{a}_i$ s.t.

$$a_1 b_1 + \cdots + a_r b_r = 1. \tag{23}$$

(**Hint:** Recall that the product of the ideals $\mathfrak{a}$ and $\mathfrak{b}$ is not $\{ab : a \in \mathfrak{a}$ and $b \in \mathfrak{b}\}$, But $\{\sum_{j=1}^{s} a_i b_i : a_i \in \mathfrak{a}$ and $b_i \in \mathfrak{b}\}$.)

Given (23), pick $x_1, \ldots, x_r \in M$ s.t. $\varphi_i(x_j) = b_j$ in $\mathfrak{a}_i$ for $j = 1, \ldots, r$, and let c be arbitrary in $\mathfrak{a}_i - \{0\}$. Since $a_j \in \mathfrak{a}_i^{-1}$, we see that $ca_j \in R$ and

$$z = (ca_1)x_1 + \cdots + (ca_r)x_r \tag{24}$$

is some element of M. We also have

$$\varphi(z) = (ca_1)\varphi(x_1) + \cdots + (ca_r)\varphi(x_r) = c(a_1 b_1 + \cdots + a_r b_r) = c$$

by (23) and the choice of the x_i. Let T_i be the pure submodule of M generated by z.

Claim: $M_i = T_i \oplus N$.

First we must show that $T_i \cap N = \{0\}$. We work in $K \otimes_R M$, so N becomes $1 \otimes_R N$ and T_i becomes $W_0 \cap 1 \otimes_R M$ where W_0 is the subspace of $K \otimes_R M$ generated by $1 \otimes z$. An arbitrary element of T_i can be written as $k \otimes z$ for some $k \in K$. Suppose $k \otimes z \in 1 \otimes_R N$, i.e. $k \otimes z = 1 \otimes n$ for some $n \in N$. We extend $\varphi : M \to \mathfrak{a}_1 \oplus \cdots \oplus \mathfrak{a}_{n-1}$ to $K \otimes_R M$ as in Proposition 3.1, i.e. $\varphi(k \otimes m) = k \otimes \varphi(m)$. Since $\varphi(n) = 0$,

$$0 = \varphi(1 \otimes n) = \varphi(k \otimes z) = k\varphi(z) = kc.$$

But c was chosen to be non-zero in $\mathfrak{a}_i$, so we must have $k = 0$, and $T_i \cap N = \{0\}$.

Now suppose m is arbitrary in M_i. Let $\varphi_i(m) = k \in \mathfrak{a}_i$. Since $a_j \in \mathfrak{a}_i^{-1}$, $ka_j \in R$ for $j = 1, \ldots, r$, and if we define w by

$$w = (ka_1)x_1 + \cdots + (ka_r)x_r, \tag{25}$$

then $w \in M_i$. If we compare (24) and (25), we see that $w = kc^{-1}z$. Restating this last equation more precisely we must write $1 \otimes w = kc^{-1} \otimes z$ since $c^{-1} \notin R$, i.e. $1 \otimes w$ is in the subspace of $K \otimes_R M$ generated by z. Since w is certainly in M by (25), $w \in T_i$. Now

$$\varphi(w) = k(a_1 b_1 + \cdots + a_r b_r) = k,$$

so $\varphi(m - w) = k - k = 0$, and we see that $(m - w) \in N$. Since $m = w + (m - w)$, we have proved the claim.

Claim: $M = (T_1 + \cdots + T_{n-1}) \oplus N$.

By the previous claim, $\varphi(T_i) = \mathfrak{a}_i$, so $\varphi(T_1 + \cdots + T_{n-1} + N) = \mathfrak{a}_1 \oplus \cdots \oplus \mathfrak{a}_{n-1}$, and we see that $M = T_1 + \cdots + T_{n-1} + N$. Thus it only remains to show that the sum is direct. If $t_i \in T_i$ and $t_1 + \cdots + t_{n-1} \in N$, then

$$\varphi(t_1) + \cdots + \varphi(t_{n-1}) = 0.$$

Now each $\varphi(t_i)$ is in $\mathfrak{a}_i$ and $\varphi : M \to \mathfrak{a}_1 \oplus \cdots \oplus \mathfrak{a}_{n-1}$, so we must have each $\varphi(t_i) = 0$. Therefore by the first claim, $t_i = 0$ for $i = 1, \ldots, n-1$, and if we set $T = T_1 + \cdots + T_{n-1}$, we are done. □

Remarks: i) Recall that a module L is *projective* if we can find a homomorphism : $L \to M$ that fills in the following commutative diagram

$$\begin{array}{ccccc} & & L & & \\ & & \downarrow f & & \\ M & \xrightarrow{p} & N & \longrightarrow & 0 \end{array}$$

whenever the row is exact.

Exercise 3.6: Let $\mathfrak{a}$ be an ideal of R. Show $\mathfrak{a}$ is a projective R-module. (**Hint:** Look at the proof of the first claim in the above theorem.)

ii) Now we know that any "reasonable" modules of rank n, say M and N, can be written as

$$M = \mathfrak{a}_1 \oplus \cdots \oplus \mathfrak{a}_n \quad \text{and} \quad N = \mathfrak{b}_1 \oplus \cdots \oplus \mathfrak{b}_n.$$

It would be nice if there were a "Krull–Schmidt type theorem" for R-modules which said that M and N were isomorphic iff the $\mathfrak{a}_i$'s were isomorphic to the $\mathfrak{b}_j$'s in some order. Unfortunately such an assertion is false. What is true is contained in the following theorem:

Theorem 3.2 : *Suppose*

$$M = \mathfrak{a}_1 \oplus \cdots \oplus \mathfrak{a}_m$$

and

$$N = \mathfrak{b}_1 \oplus \cdots \oplus \mathfrak{b}_n$$

where $\mathfrak{a}_i$ and $\mathfrak{b}_j$ are ideals in R for $i = 1, \ldots, m$ and $j_1, \ldots, n$. Then M and N are isomorphic iff $m = n$ and the products $\mathfrak{a}_1\mathfrak{a}_2 \cdots \mathfrak{a}_m$ and $\mathfrak{b}_1\mathfrak{b}_2 \cdots \mathfrak{b}_n$ are in the same ideal class.

Proof: Suppose $\varphi : M \to N$ is an isomorphism. By tensoring with K and looking at dimensions, we can easily see that $m = n$. Now for each i pick a non-zero $x_i \in \mathfrak{a}_i$ and for each j pick a non-zero $y_j \in \mathfrak{b}_j$. Replace the R-module $\mathfrak{a}_i$ (respectively $\mathfrak{b}_j$) by the R-module $M_i = x_i^{-1}\mathfrak{a}_i$ (respectively $N_i = y_j^{-1}\mathfrak{b}_j$). Since we haven't changed any isomorphism classes, we have $M \approx M_1 \oplus \cdots \oplus M_m$ and $N \approx N_1 \oplus \cdots \oplus N_n$. Since we are in K, we can still multiply to get modules $M_1M_2 \cdots M_m$ and $N_1N_2 \cdots N_n$, i.e. we form the products in the same way we form the products of ideals.

Exercise 3.7: i) Show that $M_1M_2\cdots M_m$ (respectively $N_1N_2\cdots N_n$) is isomorphic to $\mathfrak{a}_1\mathfrak{a}_2\cdots\mathfrak{a}_m$ (respectively $\mathfrak{b}_1\mathfrak{b}_2\cdots\mathfrak{b}_n$).

ii) Show that φ extends to a K-linear isomorphism $\varphi: M \to N$. (**Hint:** Think about the proof of Proposition 2.3.)

For each i, we have $1 \in M_i$. (That's the point of multiplying by x_i^{-1}.) Hence we can write

$$\varphi(1) = a_{i1} \oplus a_{i2} \oplus \cdots \oplus a_{in}$$

where $a_{ik} \in N_k$. Hence if m_i is arbitrary in M_i ($\subset K$), we see that

$$\varphi(m_i) = \varphi(m_i \cdot 1) = m_i\varphi(1) = m_ia_{i1} \oplus m_ia_{i2} \oplus \cdots \oplus m_ia_{in},$$

and φ maps M_i onto $\oplus_{k=1}^n a_{ik}M_i$ and, finally, φ maps M onto $\oplus_{i=1}^m \oplus_{k=1}^n a_{ik}M_i$. From this we can see that

$$N_k = a_{1k}M_1 \oplus a_{2k}M_2 \oplus \cdots \oplus a_{mk}M_m. \tag{26}$$

Therefore the product $N_1N_2\cdots N_n$ contains all products of the form

$$a_{1k_1}a_{2k_2}\cdots a_{mk_m} \cdot M_1M_2\cdots M_m$$

where $(k_1, k_2, \ldots, k_m)$ is any permutation of $(1, 2, \ldots, m)$. Let $d = \det(a_{ik})$. Then d is a sum of products of the form $\pm a_{1k_1}a_{2k_2}\cdots a_{mk_m}$, so we get

$$N_1N_2\cdots N_n \supset dM_1M_2\cdots M_m.$$

Exercise 3.8: Show $M_1M_2\cdots M_m \supset d^{-1}N_1N_2\cdots N_n$. (**Hint:** Do the above proof starting with $1 \in N_k$, and looking at $\varphi^{-1}(1)$, etc.)

Therefore $N_1N_2\cdots N_n = dM_1M_2\cdots M_m$, which shows these products are in the same ideal class.

For the other half of the proof, we use induction and conclude that it suffices to show that if $\mathfrak{a}$ and $\mathfrak{b}$ are ideals in R, then

$$\mathfrak{a} \oplus \mathfrak{b} \approx R \oplus \mathfrak{a}\mathfrak{b}.$$

We use the following:

Lemma 3.4: *Let $\mathfrak{a}$ and $\mathfrak{b}$ be ideals of R. Then there are ideals $\mathfrak{a}'$ equivalent to $\mathfrak{a}$ and $\mathfrak{b}'$ equivalent to $\mathfrak{b}$ s.t. $\mathfrak{a}'$ are $\mathfrak{b}'$ are relatively prime.*

Assuming the lemma for a bit, we finish the proof of the theorem. Since we are only interested in isomorphism classes, we can use the lemma and assume that $\mathfrak{a} \oplus \mathfrak{b} = R$.

Now we can pick $a_0 \in \mathfrak{a}$ and $b_0 \in \mathfrak{b}$ s.t. $a_0 - b_0 = 1$. Define $\varphi : \mathfrak{a} \oplus \mathfrak{b} \to R \oplus \mathfrak{a}\mathfrak{b}$ by

$$\varphi(a, b) = (a + b, ba_0 + ab_0).$$

It is trivial to see that φ is a homomorphism of R-modules. If $\varphi(a, b) = 0$, we get first $b = -a$, and then $ba_0 - bb_0 = 0$, so $b(a_0 - b_0) = b \cdot 1 = b = 0 = a$. Thus φ is injective.

Let $x \in R$ and $y \in \mathfrak{a}\mathfrak{b}$. Put $a = a_0 x - y$ and $b = x - a = y - b_0 x$. Since $\mathfrak{a}$ and $\mathfrak{b}$ are ideals, we have $a \in \mathfrak{a}$ and $b \in \mathfrak{b}$. Now we do the following obvious computation:

$$\begin{aligned} \varphi(a, b) &= (a + b, ba_0 + ab_0) \\ &= (a + x - a, (y - b_0 x)a_0 + (a_0 x - y)b_0) \\ &= (x, y(a_0 - b_0)) \\ &= (x, y), \end{aligned}$$

so φ is surjective. □

Now we prove Lemma 3.4.

Proof: Consider $\mathfrak{a}^{-1} \subset K$. By clearing denominators, we can find $d \in R$ s.t. $d\mathfrak{a}^{-1} \subset R$. Now use Theorem 2.5 to find an ideal $\mathfrak{c}$ which is relatively prime to $\mathfrak{b}$ and s.t. $\mathfrak{c}(d\mathfrak{a}^{-1})$ is principal, say $\mathfrak{c}(d\mathfrak{a}^{-1}) = aR$ for some $a \in R$. This says that $\mathfrak{c} = (da)\mathfrak{a}$. Thus we can take $\mathfrak{b}' = \mathfrak{b}$ and $\mathfrak{a}' = \mathfrak{c}$, and these ideals will satisfy the lemma. □

An immediate consequence of Theorem 3.2 is the following:

Corollary 3.1: *Suppose $\mathfrak{a}_1, \ldots, \mathfrak{a}_n$ are ideals. Then $\mathfrak{a}_1 \oplus \cdots \oplus \mathfrak{a}_n$ is isomorphic to $R^{n-1} \oplus \mathfrak{a}_1 \cdots \mathfrak{a}_n$ where R^{n-1} is the direct sum of R with itself $n - 1$ times.*

4. Modules over Groups of Prime Order

Continuing the notation of the previous section, p will be a prime number, ς a primitive (i.e. $\varsigma \neq 1$) pth root of 1, R the ring of cyclotomic integers $\mathbb{Z}[\varsigma]$, and K the cyclotomic field $\mathbb{Q}(\varsigma)$. Φ is the group $\mathbb{Z}_p = \mathbb{Z}/p\mathbb{Z}$. The aim of this section is to classify all modules over Φ which are $\mathbb{Z}$-finitely generated and $\mathbb{Z}$-torsionfree, i.e. finitely generated and torsionfree as abelian groups. References for this section are Section 74 of [28] and [72]. Let M be such a Φ-module.

Recall that $\Sigma \in \mathbb{Z}[\Phi]$ is the element

$$\Sigma = 1 + g + g^2 + \cdots + g^{p-1},$$

where g is a generator of Φ. Set

$$M_\Sigma = \{m \in M : \Sigma \cdot m = 0\}.$$

M_Σ is a Φ-submodule of M.

Exercise 4.1: Show that M_Σ is a pure $\mathbb{Z}$-submodule of M. Thus as an abelian group, $M = M_\Sigma \oplus X$ for some $\mathbb{Z}$-submodule X of M. (**Hint:** You must show that for $a \in M$ and $m \in M$, if $am \in M_\Sigma$, then either $a = 0$ or $m \in M_\Sigma$. For this, compute $\Sigma \cdot am$.)

By construction, $\Sigma \cdot m = 0 \quad \forall m \in M_\Sigma$, so M_Σ is naturally a module over the quotient ring $\mathbb{Z}[\Phi]/(\Sigma)$ where (Σ) is the principal ideal $\Sigma \cdot \mathbb{Z}[\Phi]$.

Exercise 4.2: Show that $\mathbb{Z}[\Phi]/(\Sigma)$ is isomorphic (as a ring) to $R = \mathbb{Z}[\varsigma]$.

Therefore M_Σ is a finitely generated R-module, and the action is given by

$$\varsigma \cdot m = g \cdot m.$$

To apply the theory of the last section, we must show that M_Σ is R-torsionfree. Let $x \in R - \{0\}$. Then $N(x) \in \mathbb{Z} - \{0\}$ by Proposition 2.4. By equation (4), $N(x) = xy$ for some $y \in R$ (y is the product of the roots other than x of the minimal polynomial of x). Hence if $xm = 0$ for some $m \in M_\Sigma$, then $xym = 0$ which says $N(x)m = 0$ so $m = 0$, and we see that M is R-torsionfree. We also have the inclusions

$$(g - 1)M \subset M_\Sigma \tag{28}$$

and

$$(\varsigma - 1)M_\Sigma \subset (g - 1)M. \tag{29}$$

It is trivial to see that (29) holds, and (28) follows since $\Sigma(g - 1) = 0$ (a fact that led to the nice resolution of $\mathbb{Z}$ for Φ in section 3 of Chapter III).

By Theorem 3.1, M_Σ is isomorphic to a direct sum of ideals of R, and by Theorem 3.2, M_Σ is determined by the number of these ideals and the ideal class of their product. Using Corollary 3.1, we can write

$$M_\Sigma = b_1 R \oplus \cdot \oplus b_{n-1} R \oplus b_n \mathfrak{a} \tag{30}$$

with $b_1, \ldots, b_n \in M_\Sigma$ and $\mathfrak{a}$ an ideal of R. Since $(g-1)M$ is a submodule of M_Σ of the same rank, we can also write

$$(g-1)M_\Sigma = b_1 \mathfrak{a}_1 \oplus \cdots \oplus b_{n-1}\mathfrak{a}_{n-1} \oplus b_n \mathfrak{a}_n \mathfrak{a}$$

where $\mathfrak{a}_1, \ldots, \mathfrak{a}_n$ are ideals of R. Equation (29) implies

$$(\varsigma - 1)R \subset \mathfrak{a}_i \subset R \qquad \text{for } i = 1, \ldots, n.$$

Exercise 4.3: Show that $(\varsigma - 1)R$ is a maximial ideal. (**Hint:** What is $N(\varsigma - 1)$?)

We therefore see that each $\mathfrak{a}_i$ is either R or $(\varsigma - 1)R$.

Exercise 4.4: Let $\mathfrak{a}$ and $\mathfrak{b}$ be ideals of R. Show that there is an R-isomorphism $\varphi : R \oplus \mathfrak{b} \to R \oplus \mathfrak{b}$ s.t. $\varphi(\mathfrak{a} \oplus \mathfrak{b}) = R \oplus \mathfrak{ab}$. (**Hint:** Use the φ defined in the second half of Theorem 3.2.)

Therefore by rearranging and using Exercise 4.4 if necessary, we can assume $\mathfrak{a}_1 = \mathfrak{a}_2 = \cdots = \mathfrak{a}_r = R$ and $\mathfrak{a}_{r+1} = \mathfrak{a}_{r+2} = \cdots = \mathfrak{a}_n = (\varsigma - 1)R$. There will be two cases, case I for $r < n$ and case II for $r = n$.

Consider the quotient module

$$N = \frac{(g-1)M}{(\varsigma - 1)M_\Sigma}.$$

In case I we have

$$N = \frac{R}{(\varsigma - 1)R} \oplus \cdots \oplus \frac{R}{(\varsigma - 1)R},$$

where there are r terms in the direct sum. For case II we get

$$N = \frac{R}{(\varsigma - 1)R} \oplus \cdots \oplus \frac{R}{(\varsigma - 1)R} \oplus \frac{\mathfrak{a}}{(\varsigma - 1)\mathfrak{a}},$$

where there are $n - 1$ terms of the form $R/(\varsigma - 1)R$.

Exercise 4.5: i) Show that $R/(\varsigma-1)R$ is isomorphic to $\mathbb{Z}_p$. (**Hint:** What is $N(\varsigma-1)$?)

ii) Show that $\mathfrak{a}/(\varsigma-1)\mathfrak{a}$ is also isomorphic to $\mathbb{Z}_p$. (**Hint:** This is harder. You want to show that

$$\frac{R}{(\varsigma-1)R}\approx\frac{\mathfrak{a}}{(\varsigma-1)\mathfrak{a}}.$$

Use Theorem 2.5 to get an ideal $\mathfrak{b}$ which is prime to $(\varsigma-1)R$ and s.t. $\mathfrak{a}\mathfrak{b}=xR$. Then look at the group homomorphism $\varphi: R\to\mathfrak{a}/(\varsigma-1)\mathfrak{a}$ defined by $\varphi(y)=xy+(\varsigma-1)\mathfrak{a}$, and show that $\ker\varphi=(\varsigma-1)R$.)

Thus in either case N is isomorphic to a vector space over $\mathbb{Z}_p$ of dimension r. Let $\beta_i\in N$ be the class of $b_i\in M_\Sigma$. We can write

$$N=\beta_1\mathbb{Z}_p\oplus\cdots\oplus\beta_r\mathbb{Z}_p.$$

From Exercise 4.1 we know that $M=M_\Sigma\oplus X$ as $\mathbb{Z}$-modules, so

$$(g-1)M=(g-1)M_\Sigma+(g-1)X=(\varsigma-1)M_\Sigma+X.$$

Since X is not a Φ-submodule of M_Σ , this sum is not direct, so we cannot conclude that $N\approx X$. We do know, however, that the elements of N are the cosets of $(\varsigma-1)M_\Sigma$ and thus each such coset contains an element of $(g-1)X$. Hence the linear map $\varphi: X\to N$ defined by

$$\varphi(x)=(g-1)x+(\varsigma-1)M_\Sigma \tag{31}$$

is a $\mathbb{Z}$-homomorphism of X *onto* N. Let $x_1,\dots,x_k$ be a $\mathbb{Z}$-basis of X and define a matrix $A=(a_{ij})$ by

$$\varphi(x_i)=\sum_{j=1}^{r}a_{ij}\beta_j.$$

A is a $k\times r$ matrix with entries in $\mathbb{Z}_p$. Since φ is surjective, the rank (over $\mathbb{Z}_p$) of A is precisely r. This shows that $k\geq r$.

Suppose we change the basis $x_1,\dots,x_k$ of X to another basis, say $y_i,\dots,y_k$, so

$$y_i=\sum_{j=1}^{k}u_{ij}x_j$$

with $u_{ij}\in\mathbb{Z}$. Thus $U=(u_{ij})$ is a $k\times k$ unimodular matrix. We have

$$\varphi(y_i)=\sum_{j,l}\underline{u}_{ij}a_{il}\beta_l$$

where $\underline{u}_{ij}$ is the reduction modulo p of u_{ij}. Hence when we change basis in X, the matrix A changes to the matrix $\underline{U}A$ where $\underline{U}=(\underline{u}_{ij})$.

Exercise 4.6: Show that it is possible to pick a basis for X so that the matrix $\underline{U}A$ has the form

$$\begin{pmatrix} c_1 & 0 & \cdots & 0 \\ 0 & c_2 & \cdots & 0 \\ \vdots & \vdots & & \vdots \\ 0 & 0 & \cdots & c_r \\ 0 & 0 & \cdots & 0 \\ \vdots & \vdots & & \vdots \\ 0 & 0 & \cdots & 0 \end{pmatrix}$$

with $c_i \in \mathbb{Z}_p - \{0\}$. (**Hint:** This is a perfectly standard problem in linear algebra. Use row and column operations and the fact that the rank is r.)

We now can assume that our basis $x_1, \ldots, x_k$ has this property. In other words we now can assume that

$$\varphi(x_i) = c_i b_i \qquad \text{for } i = 1, 2, \ldots, r$$

and

$$\varphi(x_i) = 0 \qquad \text{for } i = r+1, \ldots, k.$$

From (31) we get

$$(g-1)x_i \equiv \overline{c}_i b_i \ (\mathrm{mod}(\varsigma - 1)M_\Sigma) \qquad \text{for } i = 1, \ldots, r$$

and

$$(g-1)x_i \equiv 0 \ (\mathrm{mod}(\varsigma - 1)M_\Sigma) \qquad \text{for } i = r+1, \ldots, k$$

where in cases I and II, $\overline{c}_i \in \mathbb{Z}$ is any integer that reduces modulo p to c_i for $i = 1, 2, \ldots, r$ and $r < n$, while in case II, $\overline{c}_n \in \mathfrak{a}$ is any element that reduces modulo $(\varsigma - 1)\mathfrak{a}$ to c_n. Thus we can pick $u_1, \ldots, u_k \in M_\Sigma$ s.t.

$$(g-1)x_i = \overline{c}_i b_i + (\varsigma - 1)u_i \qquad \text{for } i = 1, \ldots, r \tag{32}$$

and

$$(g-1)x_i = (\varsigma - 1)u_i \qquad \text{for } i = r+1, \ldots, k. \tag{33}$$

Define $y_1, \ldots, y_k \in M$ by

$$y_i = x_i - u_i \qquad \text{for } i = 1, \ldots k. \tag{34}$$

Recall that $M = M_\Sigma \oplus X$ as $\mathbb{Z}$-modules and that $\{x_1, \ldots, x_k\}$ is a basis for X and that $u_1, \ldots u_k \in M_\Sigma$. Hence although $\{y_1, \ldots, y_k\}$ is not contained in X, the $\mathbb{Z}$-module generated by the y_i's is a complement of M_Σ in M. Thus we can write

$$M = M_\Sigma \oplus y_1\mathbb{Z} \oplus \cdots \oplus y_k\mathbb{Z} \tag{35}$$

as $\mathbb{Z}$-modules. Using (32), (33), and (34) we get

$$g \cdot y_i = \bar{c}_i b_i + y_i \qquad \text{for } i = 1, \ldots, r \tag{36}$$

and

$$g \cdot y_i = y_i \qquad \text{for } i = r+1, \ldots, k. \tag{37}$$

Now use (30) to expand the term M_Σ in (35) and rearrange to yield

$$M = (b_1R \oplus y_1\mathbb{Z}) \oplus \cdots \oplus (b_rR \oplus y_r\mathbb{Z}) \oplus b_{r+1}R \oplus \cdots \oplus b_{n-1}R \oplus b_n\mathfrak{a} \oplus y_{r+1}\mathbb{Z} \oplus y_k\mathbb{Z} \tag{38}$$

in case I. In case II we get

$$M = (b_1R \oplus y_1\mathbb{Z}) \oplus \cdots \oplus (b_{n-1}R \oplus y_{n-1}\mathbb{Z}) \oplus (b_n\mathfrak{a} \oplus y_n\mathbb{Z}). \tag{39}$$

Ostensibly the decompositions (38) and (39) are only decompositions over $\mathbb{Z}$, however, we can see from (36) and (37) that the terms $(b_iR \oplus y_i\mathbb{Z})$ and $(b_n\mathfrak{a} \oplus y_n\mathbb{Z})$ are actually Φ-modules, so, *mirabile dictu*, if we regard the sums in parenthesis as single Φ-modules, (38) and (39) are, in fact, decompositions of M as a direct sum of Φ-submodules. This is, of course, the reason we arranged the sums in precisely this fashion.

Let us examine the Φ-modules $(b_iR \oplus y_i\mathbb{Z})$ and $(b_n\mathfrak{a} \oplus y_n\mathbb{Z})$. They are both constructed in the following manner: Let $\mathfrak{a}$ be an ideal in R and fix an element $a_0 \in \mathfrak{a}$. Define a Φ-module structure on $\mathfrak{a} \oplus \mathbb{Z}$ by

$$g \cdot (a, m) = (\varsigma a + ma_0, m) \tag{40}$$

for $a \in \mathfrak{a}$ and $m \in \mathbb{Z}$.

Exercise 4.7: Show that g^p acts as the identity on $\mathfrak{a} \oplus \mathbb{Z}$, so (40) really does define an action of Φ.

We denote this module by $\beta(\mathfrak{a}, a_0)$. Now the Φ-modules $(b_iR \oplus y_i\mathbb{Z})$ and $(b_n\mathfrak{a} \oplus y_n\mathbb{Z})$ can be written

$$(b_iR \oplus y_i\mathbb{Z}) \approx \beta(R, c_i) \qquad \text{for } i = 1, \ldots, r$$

and

$$(b_n \mathfrak{a} \oplus y_n \mathbb{Z}) \approx \beta(\mathfrak{a}, c_n).$$

Lemma 4.1: *Suppose $c \in \mathbb{Z}$ and $p \nmid c$. Then $\beta(\mathfrak{a}, a_0) \approx \beta(\mathfrak{a}, ca_0)$.*

Proof: By definition we can write

$$\beta(\mathfrak{a}, a_0) = \mathfrak{a} \oplus y_1 \mathbb{Z},$$
$$ga = \varsigma a \qquad \text{for } a \in \mathfrak{a},$$
$$gy_1 = a_0 + y_1,$$

and

$$\beta(\mathfrak{a}, ca_0) = \mathfrak{a} \oplus y_2 \mathbb{Z},$$
$$ga = \varsigma a \qquad \text{for } a \in \mathfrak{a},$$
$$gy_2 = ca_0 + y_2,$$

where we have used y_1 and y_2 to distinguish the different generators of the free abelian subgroups.

Let

$$\theta = \frac{\varsigma^c - 1}{\varsigma - 1} = \sum_{i=0}^{c-1} \varsigma^i \in R.$$

Since $p \nmid c$, we can choose $d \in \mathbb{Z}$ s.t. $cd \equiv 1 \pmod{p}$.

Exercise 4.8: Show that

$$\frac{\varsigma^{cd} - 1}{\varsigma - 1} = \theta^{-1},$$

so θ is a unit in R.

If $c > 0$, we have

$$\theta - c = \sum_{i=0}^{c-1} (\varsigma^i - 1) \equiv 0 \ (\mathrm{mod}(\varsigma - 1)R) \tag{41}$$

since $(\varsigma - 1) | (\varsigma^i - 1)$ for $i = 1, \ldots, c$. Equation (41) also holds if $c < 0$ because we can replace c by $c + kp$ and $p \in (\varsigma - 1)R$ by (7). Hence $(\theta - c)a_0 \equiv 0 \ (\mathrm{mod}(\varsigma - 1)\mathfrak{a})$. This means that there is an element $\omega \in \mathfrak{a}$ s.t.

$$(\theta - c)a_0 = (\varsigma - 1)\omega. \tag{42}$$

We now define a map $\varphi : \beta(\mathfrak{a}, a_0) \to \beta(\mathfrak{a}, ca_0)$ by $\varphi(a, my_1) = (\theta a + m\omega, my_2)$. The map φ is clearly an isomorphism of abelian groups, so if we can show that φ is Φ-linear, we will be done. Trivially, $g \cdot \varphi(a,0) = \varphi(ga,0)$ $\forall a \in \mathfrak{a}$. We also have

$$g \cdot \varphi(0,y) = g \cdot (\omega, y_2) = (\varsigma\omega + ca_0, y_2) = (\omega + \theta a_0, y_2)$$

by (42) and

$$\varphi(0, g \cdot y) = \varphi(a_0, y_1) = (\theta a_0 + \omega, y_2).$$

□

Exercise 4.9: i) Show that $(R,c) \approx (R,1)$ if $c \in \mathbb{Z}$ and $p \not| c$.

ii) Show that $(R,1)$ is isomorphic to the Φ-module $\mathbb{Z}[\Phi]$. (**Hint:** Try $y \mapsto 1-g$.)

Thus in the decompositions (38) and (39) we can replace each $b_i R \oplus y_i \mathbb{Z}$ by $b_i\mathbb{Z}[\Phi]$. We are left with examining the module $\beta(\mathfrak{a}, c_n) = \mathfrak{a} \oplus y_n\mathbb{Z}$. In this module we have $g \cdot y_n = \overline{c}_n b_n + y_n$ where $\overline{c}_n \in \mathfrak{a}$ reduces to c_n modulo $(\varsigma - 1)\mathfrak{a}$. Since $c_n \neq 0$ in $\mathfrak{a}/(\varsigma - 1)\mathfrak{a}$, $\overline{c}_n$ is not in $(\varsigma - 1)\mathfrak{a}$.

We now want to fix for all time, now and forevermore, some element $a_0 \in \mathfrak{a}$ s.t. $a_0 \notin (\varsigma - 1)\mathfrak{a}$. Since $a_0 \notin (\varsigma - 1)\mathfrak{a}$, some multiple of a_0 is in the same residue class as $\overline{c}_n$ modulo $(\varsigma - 1)\mathfrak{a}$, i.e. $\exists h \in \mathbb{Z}$ with $p \not| h$ s.t.

$$\overline{c}_n \equiv ha_0 \pmod{(\varsigma - 1)\mathfrak{a}}.$$

Now $\overline{c}_n$ is only determined modulo $(\varsigma - 1)\mathfrak{a}$, so we may as well assume $\overline{c}_n = ha_0$. Lemma 4.1 tells us that

$$\beta(\mathfrak{a}, c_n) = \beta(\mathfrak{a}, ha_0) \approx \beta(\mathfrak{a}, a_0).$$

Finally we can write $\beta(\mathfrak{a}, a_0)$ simply as $\beta(\mathfrak{a})$ because we have fixed a_0 forever.

We are getting into position to pull together all the results of this section in a big theorem. Let's go back to the decomposition (38). In case I, we have shown that every finitely generated torsionfree Φ-module M can be written as the direct sum of a certain number (r in (38)) of group rings, $\mathbb{Z}[\Phi] = \beta(R,1)$, a certain number ($n-r-1$ in (38)) of the rings $R(= \mathbb{Z}[\varsigma])$, an ideal $\mathfrak{a}$ of R, and a certain number ($k-r$ in (38)) of $\mathbb{Z}$'s on which Φ acts trivially, i.e.

$$M \approx \mathbb{Z}[\Phi] \oplus \cdots \oplus \mathbb{Z}[\Phi] \oplus \mathbb{Z}[\varsigma] \oplus \cdots \oplus \mathbb{Z}[\varsigma] \oplus \mathfrak{a} \oplus \mathbb{Z} \oplus \cdots \oplus \mathbb{Z}. \tag{43}$$

Similarly in case II, M can be written

$$M \approx \mathbb{Z}[\Phi] \oplus \cdots \oplus \mathbb{Z}[\Phi] \oplus \beta(\mathfrak{a}) \oplus \mathbb{Z} \oplus \cdots \oplus \mathbb{Z}. \tag{44}$$

Keeping these decompositions in mind, we state the promised big theorem.

Theorem 4.1 (Diederichsen–Reiner): *Let M be a Φ-module which is finitely generated and torsionfree as an abelian group. Let M_Σ be the submodule of M defined by $M_\Sigma = \{m \in M : \Sigma \cdot m = 0\}$. Let n be the rank of M_Σ as an R-module, k the rank of M/M_Σ as a free abelian group, and r the dimension of $(g-1)M/(\varsigma-1)M_\Sigma$ as a vector space over $\mathbb{Z}_p$. Write M_Σ as $R \oplus \cdots \oplus R \oplus \mathfrak{a}$ where $\mathfrak{a}$ is an ideal of R. Then the isomorphism class of M as an Φ-module is determined by the integers $n, k,,$ and r and the ideal class of $\mathfrak{a}$.*

Conversely, if $n, k,,$ and r are integers with $k \geq r \geq 0$ and $n \geq r \geq 0$, and $[\mathfrak{a}]$ is some ideal class, then the Φ-module M constructed by (38) for $r < n$ and by (39) for $r = n$, is a Φ-module with the invariants n, k, r, and $[\mathfrak{a}]$.

Remark: Roughly speaking, there are three types of Φ-modules out of which all Φ-modules are built. The first type (denoted by "1" in [19]) is $\mathbb{Z}$ with trivial action. The second type (denoted by "α" in [19]) is $R = \mathbb{Z}[\varsigma]$ or an ideal of R. The third type (denoted by "β" in [19]) is $\mathbb{Z}[\Phi]$ or $\beta(\mathfrak{a})$ for some ideal $\mathfrak{a}$. R and $\mathfrak{a}$ are very much alike, and so are $\mathbb{Z}[\Phi]$ and $\beta(\mathfrak{a})$. One way of saying this is that they are the same globally and only differ locally. This means that if you tensor them with something of characteristic p (say the local ring of $\mathbb{Z}$ at p), they become identical. See [19] for details.

We can make more precise the notion of building up Φ-modules from the three basic types.

Definition 4.1: We say a module M over a ring $\mathcal{R}$ is *indecomposable* if M cannot be written as a (non-trivial) direct sum.

Exercise 4.10: Show that $\mathbb{Z}, \mathfrak{a}$, and $\beta(\mathfrak{a})$ (including the cases $\mathfrak{a} = R$) are the only indecomposable Φ-modules.

Thus we see that there are precisely $2h_p+1$ indecomposable Φ-modules where h_p is the class number of the cyclotomic ring (or field). Clearly every Φ-module (finitely generated and torsionfree over $\mathbb{Z}$) is a direct sum of indecomposable Φ-modules. It is important to remember that this decomposition will *not* be unique.

Remark: The problem of finding an analogue of Theorem 3.1 for groups other than $\mathbb{Z}_p$ is very difficult. If G is cyclic of arbitrary order, for example, the corresponding result is unknown at the time of writing, although if the order is squarefree, the indecomposable modules are known . The indecomposable representations of the dihedral group of order $2p$ where p is an odd prime and, more generally, the indecomposable representations of the metacyclic group with a (normal) subgroup of order p and of index q (with p and q prime and unequal) are known. All of these groups have the property that the number of indecomposables is finite. It is also known that the number of indecomposables for a cyclic group of order p^2 (p prime) is finite although these have not been classified. The definitive result on the finiteness of the number of indecomposables is due to A. Jones ([48]).

Theorem 4.2 (Jones): *A group G has a finite number of indecomposable G-modules iff for each prime p each p-Sylow subgroup of G is cyclic of order at most p^2.*

A reference for all of these results is [29], especially around page 753.

A case in which the number of indecomposables of G is infinite, and they have all been classified, is the case $G = \mathbb{Z}_2 \oplus \mathbb{Z}_2$, the Klein 4-group. This result is due to Nazarova although her paper lacks complete proofs and is extremely laconic. Again [29] is the place to look for more results and for more references to more results.

Remark: If G is any group, one can make the collection of (isomorphism classes of) G-modules into a ring by setting

$$M + N = M \oplus N$$

and

$$M \cdot N = M \otimes_{\mathbb{Z}} N$$

where we use the diagonal action to make $M \otimes_{\mathbb{Z}} N$ into a G-module. This ring is called the *integral representation ring* of G and is denoted by $R(G;\mathbb{Z})$. It is an interesting problem to determine the multiplication in this ring. For example, for $G = \mathbb{Z}_p$ it turns out that $\beta \otimes M = (\mathrm{rk}M)\beta$ if β is the module $\mathbb{Z}[\Phi]$ and M is arbitrary. An even more curious result is $\alpha \otimes \alpha = 1 \oplus (p-2)\beta$ where $\alpha = \mathbb{Z}[\varsigma] = R$ and $1 = \varsigma$ with trivial action. For the rest of the structure of this representation ring and also the structure of $R(D_{2p};\mathbb{Z})$, see [22].

Remark: Another interesting construction related to the representation ring is the *projective class group.* This group measures the difference between projective modules and free modules. Let $\mathcal{R}$ be any ring. Define an equivalence on the set of (isomorphism classes of) projective $\mathcal{R}$-modules by $P_1 \sim P_2$ iff there are free modules F_1 and F_2 s.t. $P_1 \oplus F_1 \approx P_2 \oplus F_2$.

Exercise 4.11: i) Show that the set of equivalence classes forms a group under the composition induced by direct sum. This group is called the *projective class group* of $\mathcal{R}$ and is denoted by $\mathcal{P}(\mathcal{R})$. (**Hint:** The only somewhat difficult part is the existence of inverses, and Proposition 3.1 of Chapter III is just what you need for that.)

ii) In the case $\mathcal{R} = \mathbb{Z}[\mathbb{Z}_p]$ with p prime, prove the following theorem of D. S. Rim ([71]):

Theorem 4.3 (Rim): $\mathcal{P}(\mathcal{R}) \approx C_p$, *the ideal class group of* $\mathbb{Z}[\varsigma]$.

(**Hint:** First show that the indecomposable projective $\mathbb{Z}_p$-modules are the modules $\beta(\mathfrak{a})$ for $\mathfrak{a}$ an ideal of $\mathbb{Z}[\varsigma]$. Then use the relation $\mathfrak{a} \oplus \mathfrak{b} \approx R \oplus \mathfrak{a}\mathfrak{b}$ to get

$$\beta(\mathfrak{a}) \oplus \beta(\mathfrak{b}) \approx \mathbb{Z}[\Phi] \oplus \beta(\mathfrak{a}\mathfrak{b}).)$$

5. The Cohomology of Modules over Groups of Prime Order

According to our scheme for classifying Bieberbach groups and flat manifolds, the next step is to find all classes in $H^2(\Phi; M)$ that give rise to torsionfree extensions. Since we are taking $\Phi = \mathbb{Z}_p$, Theorem 2.1 says that a class in $H^2(\Phi; M)$ gives rise to a torsionfree extension iff it is non-zero. We are therefore left with computing $H^2(\Phi; M)$. By Proposition 5.2, we reduce to the case of M indecomposable, and by Proposition 4.2, it suffices to find $\ker\Delta$ and $\Sigma \cdot M$ where $\Delta, \Sigma \in \mathbb{Z}[\Phi]$ are the elements $\Delta = g - 1$ and $\Sigma = 1 + g + g^2 + \cdots + g^{p-1}$.

Suppose M is $\mathbb{Z}$ with trivial Φ-action. Then $\ker\Delta = \mathbb{Z}$ and $\Sigma \cdot \mathbb{Z} = p\mathbb{Z}$, so $H^2(\Phi; M) \approx \mathbb{Z}_p$.

Suppose M is an ideal of R. Then $\ker\Delta = \{0\}$ since $\Delta a = 0 \Rightarrow \varsigma a = a \Rightarrow a = 0$. Since $1 + \varsigma + \varsigma^2 + \ldots + \varsigma^{p-1} = 0$, $\Sigma a = 0$ as it must. So $H^2(\Phi; \mathfrak{a}) = 0$.

So we are left with the case $M = \beta(\mathfrak{a})$. Recall that $\beta(\mathfrak{a}) = \mathfrak{a} \oplus \mathbb{Z}$ and

$$g \cdot (a, n) = (\varsigma a + n a_0, n)$$

where a_0 is a fixed element of $\mathfrak{a}$ not in $(\varsigma-1)\mathfrak{a}$. Suppose $(a,n) \in \ker\Delta$. Then

$$(a,n) = (\varsigma a + na_0, n)$$

or

$$(\varsigma - 1)a = -na_0$$

or

$$na_0 \in (\varsigma - 1)\mathfrak{a}.$$

Let $p : R \to r/(\varsigma-1)R$ be the projection. Since $a_0 \in \mathfrak{a}$ and $a_0 \notin (\varsigma-1)\mathfrak{a}$, $p(a_0)$ generates $\mathfrak{a}/(\varsigma-1)\mathfrak{a}$ because we know by part ii) of Exercise 4.5 that $\mathfrak{a}/(\varsigma-1)\mathfrak{a}$ is $\mathbb{Z}_p$. Now $p(na_0) = p(n)\cdot p(a_0) = 0$, so we must have $p(n) = 0$ which implies $n \in (\varsigma-1)\mathfrak{a}$. Clearly $n \in \mathbb{Z}$, so $n \in (\varsigma-1)R \cap \mathbb{Z}$. By part i) of Lemma 2.2 and Theorem 2.2, $(\varsigma-1)R \cap \mathbb{Z} = p\mathbb{Z}$. Hence $n = tp$ for $t \in \mathbb{Z}$, and we get that $(a,n) \in \ker\Delta \Rightarrow n = tp$ for some $t \in \mathbb{Z}$, and furthermore $(\varsigma-1)a = -tpa_0$. But it is trivial to see that $(a,tp) \in \ker\Delta$ if $(\varsigma-1)a = -tpa_0$, so

$$\ker\Delta = \{(a,tp) \in \beta(\mathfrak{a}) : t \in \mathbb{Z} \text{ and } (\varsigma-1)a = -tpa_0\}.$$

Notice that the map $a \mapsto (\varsigma-1)a$ is injective, so if we know t, we know $-tpa_0 = (\varsigma-1)a$, and we know a. Thus we have proved the following:

Lemma 5.1: *KerΔ is isomorphic to $\mathbb{Z}$ where $t \in \mathbb{Z}$ corresponds to the element $((\varsigma-1)^{-1}(tpa_0), tp) \in \beta(\mathfrak{a})$.*

Now we look at $\Sigma\cdot\beta$.

$$\Sigma\cdot(a,n) = \Sigma\cdot(a,0) + \Sigma\cdot(0,n) = (0,0) + (0, n\Sigma\cdot 1),$$

and

$$g(0,1) = (a_0, 1),$$
$$g^2(0,1) = g(a_0,1) = (\varsigma a_0 + a_0, 1),$$

so

$$g^j\cdot(0,1) = \left(\sum_{k=0}^{j-1} \varsigma^k \cdot a_0, 1\right),$$

and

$$\Sigma \cdot (0,1) = \left(\sum_{j=0}^{p-1}\sum_{k=0}^{j-1} \varsigma^k \cdot a_0, p\right)$$
$$= \left(\sum_{j=0}^{p-1}(p-j)\varsigma^j \cdot a_0, p\right).$$

Therefore $(x,n) \in \Sigma \cdot \beta(\mathfrak{a})$ iff $n = tp$ and

$$x = t\sum_{j=0}^{p-1}(p-j)\varsigma^j \cdot a_0.$$

But

$$(\varsigma - 1)\sum_{j=0}^{p-1}(p-j)\varsigma^j = (\varsigma-1)(p+(p-1)\varsigma+\cdots+\varsigma^{p-1})$$
$$= p\varsigma + (p-1)\varsigma^2 + \cdots + \varsigma^p - p - (p-1)\varsigma - \cdots - \varsigma^{p-1}$$
$$= p - (1+\varsigma+\varsigma^2+\cdots+\varsigma^{p-1})$$
$$= p.$$

Therefore

$$x = t\sum_{j=0}^{p-1}(p-j)\varsigma^j \cdot a_0 = (\varsigma-1)^{-1}(tpa_0),$$

and the elements of $\Sigma \cdot \beta(\mathfrak{a})$ are precisely the elements of $\ker\Delta$, and we have proved the following:

Theorem 5.1 : *Let M be a Φ-module which is the direct sum of A copies of $\mathbb{Z}$ with trivial action (or "1's"), B ideals of R (or "α's"), and C copies of modules of the form $\beta(\mathfrak{a})$ (or "β's"). Then*

$$H^2(\Phi; M) \approx \mathbb{Z}_p^A.$$

Exercise 5.1: Show that if M is the Φ-module of Theorem 5.1, then

$$H^0(\Phi; M) \approx \mathbb{Z}^{A+C},$$
$$H^{2n}(\Phi; M) \approx \mathbb{Z}_p^A,$$

and

$$H^{2n+1}(\Phi; M) \approx \mathbb{Z}_p^B.$$

Remark: The invariants A, B, and C of Theorem 5.1 are related to Reiner's invariants n, k, and r of Theorem 4.1 by the following equations:

$$A = k - r,$$

$$B = n - r,$$

and

$$C = r,$$

so we can use either set of invariants to describe a Φ-module. We sometimes use the notation $M(A, B, C; \mathfrak{a})$ or $M(n, k, r; \mathfrak{a})$ to denote the Φ-module specified by these invariants. The inequalities $k \geq r \geq 0$ and $n \geq r \geq 0$ are equivalent to the inequalities $A, B, C \geq 0$ which, given the definitions of A, B, and C, make sense. Finally notice that by Exercise 5.1, the "global invariants" of a Φ-module, namely A, B, and C, are determined by $H^i(\Phi; M)$ (in fact by H^0, H^1, and H^2), but the "local invariant," $\mathfrak{a}$, is not determined by cohomology. In the case of the dihedral group D_{2p}, the global invariants are *not* determined by cohomology although the groups $H^i(\Phi; M)$ do contain a lot of global information (see [22]).

6. The Classification Theorem

We are now ready to complete the classification of Bieberbach groups whose holonomy group has prime order. It turns out that Theorem 2.2 of Chapter III suffices in this case. We now know all the Φ-modules for $\Phi = \mathbb{Z}_p$ with p prime. Furthermore, we know $H^2(\Phi; M)$, and it follows immediately from Theorem 2.1 of Chapter III that a class $\alpha \in H^2(\Phi; M)$ corresponds to a torsionfree extension (and hence a Bieberbach group) iff $\alpha \neq 0$.

By Theorem 2.2 of Chapter III, we are reduced to considering pairs (M_1, γ_1) and (M_2, γ_2) with M_1 and M_2 Φ-modules and $\gamma_1, \gamma_2 \in H^2(\Phi; M)$. We want to find out if there is a semi-linear isomorphism $(f, A) : M_1 \to M_2$ s.t.

$$f_*(\gamma_1) = A^*(\gamma_2).$$

If there is, then (M_1, γ_1) and (M_2, γ_2) determine the same (i.e. isomorphic) Bieberbach groups; if not, they do not.

We can break this problem into two pieces. If $(f, A) : M_1 \to M_2$ is a semi-linear isomorphism, then f must be an isomorphism from M_1 to $A^{-1}(M_2)$. So we can first try to find an isomorphism f from M_1 to M_2 with $f_*(\gamma_1) = \gamma_2$ and then see how M_2 is related to $A^{-1}(M_2)$.

In this section we need to call $A^{-1}(M)$, $A^*(M)$ because we will need $A^{-1}$for something else, namely the inverse to the function A. Also we will write $M = A1 \oplus B\alpha \oplus C\beta$ to mean that M is the module of Theorem 5.1. Notice that we are using "A" in two different ways here, so be careful.

Recall that Φ has order p with p prime. We assume that p is not 2 or 3 as these cases must be handled separately and are trivial anyway. The "interested reader" may want to work out the details of these cases as an exercise. Our first result is the following:

Theorem 6.1 : *Let M be a Φ-module and γ and γ' non-zero classes in $H^2(\Phi; M)$. Then there is a Φ-automophism $f : M \to M$ with $f_*(\gamma) = \gamma'$ iff either*

$$i) M \neq 1 \oplus B\alpha,$$

or

$$ii) M = 1 \oplus B\alpha \quad \text{and } \gamma' = \pm\gamma$$

where $B \in \mathbb{Z}$.

Proof: If $\gamma \neq 0$, then $H^2(\Phi; M)$ can't be zero, so by Theorem 5.1, the number A is larger than 0. Then we can write $M = M_1 \oplus M_2$ where $M_1 = A \cdot 1$, i.e. M_1 is the largest direct summand of M on which Φ acts trivially. Hence $H^2(\Phi; M) = H^2(\Phi; M_1)$. Now it will suffice to find an appropriate $\mathbb{Z}$-automorphism of M_1, since the $\mathbb{Z}$-automorphisms of M_1 are also the Φ-automorphisms of M_1, and they can be extended trivially to Φ-automorphisms of all of M. (Notice that possibly not all of the Φ-automorphisms of M will be obtained in this way, i.e. a"1" can be mapped into the submodule $(\Sigma) \subset \beta = \mathbb{Z}[\Phi]$).

Now $H^2(\Phi; M_1) \approx M_1/pM_1 \approx (\mathbb{Z}_p)^A$ which is merely a vector space of dimension A over $\mathbb{Z}_p$.

Exercise 6.1: i) It is a standard fact that if V is a finite dimensional vector space, then $GL(V)$ is generated by elementary matrices, i.e. matrices with plus or minus one's along the main diagonal, zero's elsewhere except for precisely one other entry which is a plus one, e.g.

$$\begin{pmatrix} 1 & 0 & 0 & 0 \\ 0 & -1 & 0 & 1 \\ 0 & 0 & 1 & 0 \\ 0 & 0 & 0 & 1 \end{pmatrix}.$$

Prove this. (**Hint:** Use elementary row operations)

ii) Show that $GL(V)$ operates transitively on $V - \{0\}$ if the dimension of V is > 1. ("Transitively" means that given any $v_1, v_2 \in V - \{0\}$, $\exists g \in GL(V)$ s.t. $g \cdot v_1 = v_2$.)

Any elementary matrix over $\mathbb{Z}$ can be considered as an elementary matrix over $\mathbb{Z}_p$. Therefore if the dimension of M_1/pM_1 (which is the number A) is greater than 1, we can map any non-zero element of M_1/pM_1 into any other non-zero element of M_1/pM_1 by picking an appropriate $\mathbb{Z}$-automorphism of M_1, and using the map that it induces on M_1/pM_1. This proves the theorem if $A > 1$.

If $A = 1$, the above argument fails because the only non-trivial $\mathbb{Z}$-automorphism of $\mathbb{Z}$ (which is M_1 in this case) is $1 \mapsto -1$ which is certainly not transitive on $(\mathbb{Z}/p\mathbb{Z}) - \{0\}$ (unless $p = 2$ or 3 which we ignore). However, if we are in case ii) where $M = 1 \oplus B\alpha$, we can get the desired result easily. Since there are no β's, every Φ-automophism of M must preserve M_1 and hence induces a $\mathbb{Z}$-automorphism of M_1 of which there is only $1 \mapsto -1$, so γ must be mapped to $-\gamma$, and we have proved the theorem in case ii).

We are left with the situation $M = 1 \oplus B\alpha \oplus C\beta$ where $B \geq 0$ and $C > 0$. We need a Φ-automophism of M which mixes up the 1's and the β's. Since $C > 0$, we can assume that $\beta(\mathfrak{a})$ is a direct summand of M where $\mathfrak{a}$ is an ideal (possibly principal) in $\mathbb{Z}[\varsigma]$. Over $\mathbb{Z}$ (i.e. as abelian groups) we have a split exact sequence

$$0 \longrightarrow \mathfrak{a} \longrightarrow \beta(\mathfrak{a}) \xrightarrow{e} \mathbb{Z} \longrightarrow 0,$$

and the elements of $\beta(\mathfrak{a})$ invariant under Φ are $\ker\Delta$. Lemma 5.1 showed that $\ker\Delta$ was isomorphic to $\mathbb{Z}$ and had as a generator

$$u = \pm\left((\varsigma - 1)^{-1} \cdot pa_0, p\right),$$

so $e(u) = \pm p$. Let's pick u so that $e(u) = +p$. Since $A = 1$, M contains a direct summand of the form $\beta(\mathfrak{a}) \oplus \mathbb{Z}$. We define the Φ-automophism of M that takes γ to γ' by defining it on $\beta(\mathfrak{a}) \oplus \mathbb{Z}$ and letting it be the identity on the rest of M. Define

$$F : \beta(\mathfrak{a}) \oplus \mathbb{Z} \to \beta(\mathfrak{a}) \oplus \mathbb{Z}$$

by

$$F(x; n) = (\eta x + rnu; se(x) + tn), \tag{45}$$

where $x \in \beta(\mathfrak{a})$ and $n \in \mathbb{Z}$. (The reason for using a ";" to delineate the coordinates will become apparent in a couple of paragraphs.) The elements $r, s, t \in \mathbb{Z}$ and $\eta \in \mathbb{Z}[\Phi]$ are to be chosen later so as to make F a Φ-automophism which will take a specified non-zero element of $H^2(\Phi; M)$ into any other such. Since $\beta(\mathfrak{a})$ is a Φ-module, the product ηx makes sense. Also note that $\mathbb{Z}[\Phi] = \beta(R) = \beta(\mathbb{Z}[\varsigma]) = \mathbb{Z}[\varsigma] \oplus \mathbb{Z}$, where the last equality holds only as abelian groups.

If we pick $\eta \in \mathbb{Z}[\varsigma] \oplus \mathbb{Z}$ so that it projects to a unit in $\mathbb{Z}[\varsigma]$, then the map F induces from $\mathfrak{a}$ to $\mathfrak{a}$ will be a Φ-automophism because (45) implies that if $n = 0$, F is just multiplication by η on the first component. Now $\mathfrak{a}$ is a submodule of $\beta(\mathfrak{a})$ and hence of $\beta(\mathfrak{a}) \oplus \mathbb{Z}$, so F will be a Φ-automophism on $\beta(\mathfrak{a}) \oplus \mathbb{Z}$ iff it is a Φ-automophism modulo $\mathfrak{a}$.

Think of $\beta(\mathfrak{a})$ as $\mathfrak{a} \oplus \mathbb{Z}$, and use coordinates $(a, m; n)$ for $\beta(\mathfrak{a}) \oplus \mathbb{Z}$ with $a \in \mathfrak{a}$ and $m, n \in \mathbb{Z}$, so

$$g \cdot (a, m; n) = (\varsigma a + m a_0, m; n),$$

and

$$F(a, m; n) = (\eta a + Q a_0 + (\varsigma - 1)^{-1} rnp a_0, \epsilon(\eta) m + rnp; sm + tn),$$

where $\epsilon : \mathbb{Z}[\Phi] \to \mathbb{Z}$ is the usual augmentation and $Q \in \mathbb{Z}$. Actually if

$$\eta = \sum_{i=0}^{p-1} k_i g^i \in \mathbb{Z}[\Phi],$$

then

$$Q = m \sum_{i=0}^{p-1} \sum_{j=0}^{i-1} k_i \varsigma^i.$$

We have

$$F(0,1;0) = (Qa_0, \epsilon(\eta); s)$$

and

$$F(0,0;1) = \left((\varsigma - 1)^{-1} rpa_0, rp; t\right).$$

Therefore modulo $\mathfrak{a}$, F has the matrix

$$T = \begin{pmatrix} \epsilon(\eta) & rp \\ s & t \end{pmatrix}. \tag{46}$$

Now we show that F is a Φ-homomorphism. Reverting to the $(x;n)$ coordinates, we see that

$$F\left(g \cdot (x;n)\right) = F(g \cdot x; n) = (\eta g x + rnu; s\epsilon(gx) + tn),$$

while

$$g \cdot F(x,n) = g \cdot (\eta x + rnu; s\epsilon(x) + tn) = (g\eta x + rngu; s\epsilon(x) + tn).$$

Since $u \in \ker\Delta$, $g \cdot u = u$, and $\epsilon(gx) = \epsilon(x)$ by definition of the augmentation ϵ.

It now suffices to show that F is bijective. For this it suffices to show $\det(T) = \pm 1$, or

$$\epsilon(\eta) \cdot t - p \cdot rs = \pm 1. \tag{47}$$

This can be solved for r and s iff $\epsilon(\eta) \cdot t \equiv \pm 1 \pmod{p}$. Recall that η was any unit in $\mathbb{Z}[\Phi]$. If we take any q prime to p, and set $\eta = 1 + g + \cdots + g^{q-1}$, Exercise 4.8 shows that η is a unit. Clearly $\epsilon(\eta) = q$. So no matter what t is, as long as it is prime to p we can pick η s.t. $\epsilon(\eta) \cdot t \equiv \pm 1 \pmod{p}$ by setting $q \equiv \pm t^{-1} \pmod{p}$ and $\eta = 1 + g + \cdots + g^{q-1}$. Hence so long as t is prime to p, we can pick η, r and s so that (47) is satisfied. Thus F is a Φ-automorphism.

What does F_* do to $H^2(\Phi; \beta(\mathfrak{a}) \oplus \mathbb{Z})$? Since $H^2(\Phi; \beta(\mathfrak{a})) = 0$, we see that $H^2(\Phi; \beta(\mathfrak{a}) \oplus \mathbb{Z}) = H^2(\Phi; \mathbb{Z})$, and we can look at the restriction of F to $\mathbb{Z}$. Now $H^2(\Phi; \mathbb{Z})$ is isomorphic to $\mathbb{Z}_p$, and $F(0;n) = (rnu; tn)$, so the map F induces on $H^2(\Phi; \mathbb{Z})$ is merely multiplication by t. Recall that t can be chosen to be anything prime to p. Therefore we can pick F s.t. F_* maps any non-zero element of $H^2(\Phi; \mathbb{Z})$ to any other non-zero element.
□

Corollary 6.1: *Let M_1 and M_2 be isomorphic Φ-modules.*

i) If M_1 is not isomorphic to $1 \oplus B\mathfrak{a}$ and $\gamma_i \in H^2(\Phi; M_i) - \{0\}$ for $i = 1, 2$, then $\exists$ a Φ-isomorphism $F : M_1 \to M_2$ s.t. $F_*(\gamma_1) = \gamma_2$.

ii) If M_1 is isomorphic to $1 \oplus B\mathfrak{a}$ (i.e. $A = 1$ and $C = 0$) and $\gamma \in H^2(\Phi; M_1) - \{0\}$, then if $f, g : M_1 \to M_2$ are Φ-isomorphisms, we must have $f_*(\gamma) = \pm g_*(\gamma)$.

This corollary is an immediate consequence of the theorem.

We now turn our attention to $A^*(M)$ where $A \in \text{Aut}(\Phi)$. Notice that $\text{Aut}(\Phi) \approx \mathbb{Z}_{p-1}$, so there is a unique element $A_{1/2} \in \text{Aut}(\Phi)$ with the property that $(A_{1/2})^2$ is the identity. Let $[\mathfrak{a}]$ denote the ideal class of an ideal $\mathfrak{a} \subset R$, and let $N(A, B, C; [\mathfrak{a}])$ be the unique (up to isomorphism) Φ-module with invariants A, B, C, and $[\mathfrak{a}]$. Thus we could, for example, write

$$N = A \cdot 1 \oplus (B-1) \cdot \alpha \oplus \mathfrak{a} \oplus C \cdot \beta$$

if $B > 0$ and

$$N = A \cdot 1 \oplus (C-1)\beta \oplus \beta(\mathfrak{a})$$

if $B = 0$. As usual, we have used the notation that "1" is the Φ-module $\mathbb{Z}$ with trivial action, α is the Φ-module$R = \mathbb{Z}[\varsigma]$, and β is the group ring $\mathbb{Z}[\Phi]$.

Let $G = \text{Aut}(\Phi)$. We can let G act on $R = \mathbb{Z}[\varsigma]$ or $K = \mathbb{Q}(\varsigma)$ by setting $A(\varsigma) = \varsigma^d$ if A is the element in G that maps $g \in \Phi$ into g^d for $0 < d < p$.

Exercise 6.2: i) Show that this definition makes A into a field automorphism which leaves $\mathbb{Q}$ fixed, i.e. G is thus a subgroup of the galois group of $\mathbb{Q}(\varsigma)$ over $\mathbb{Q}$. Then show that G *is* this galois group.

ii) Show that $A_{1/2}$ acts as complex conjugation.

iii) Since G acts on R, it acts on ideals of R. Show that if $\mathfrak{a}$ and $\mathfrak{b}$ are in the same class, then $A \cdot \mathfrak{a}$ and $A \cdot \mathfrak{b}$ are in the same class for any $A \in G$.

We see then that G acts on C_p, the ideal class group of R.

Theorem 6.2 : *$N(A, B, C; [\mathfrak{a}])$ will be semi-linearly isomorphic to $N(A', B', C'; [\mathfrak{a}'])$ iff $A = A', B = B', C = C'$, and $A([\mathfrak{a}]) = [\mathfrak{a}']$ for some $A \in G$.*

Proof: Let $N = N(A, B, C; [\mathfrak{a}])$ and $N' = N(A, B, C; [\mathfrak{a}'])$. N and N' are semi-linearly isomorphic iff N is ismorphic to $A^*(N')$ for some $A \in G$.

It then follows that it suffices to show that

$$A^* \left(N(A, B, C; [\mathfrak{a}])\right) = N(A, B, C; A^{-1}[\mathfrak{a}]) \tag{48}$$

where A^{-1} is the inverse of A in G. (This is why we changed notation to A^*.) To prove (48), it suffices to show that

$$A^*(\mathfrak{a}) \approx A^{-1} \cdot \mathfrak{a}, \tag{49}$$

and

$$A^* \left(\beta(\mathfrak{a})\right) \approx \beta\left(A^{-1} \cdot \mathfrak{a}\right). \tag{50}$$

Suppose $A(g) = g^d$ for some d between 0 and p. Consider $A^*(\mathfrak{a})$ and the map $A^{-1} : A^*(\mathfrak{a}) \to A^{-1} \cdot \mathfrak{a}$ defined by $a \mapsto A^{-1}(a)$. We have

$$\begin{aligned} A^{-1}(g \star a) &= A^{-1}(A(g) \cdot a) \\ &= A^{-1}(\varsigma^d a) \\ &= A^{-1}(\varsigma^d) A^{-1}(a) \\ &= \varsigma A^{-1}(a) \\ &= g \cdot A^{-1}(a) \end{aligned}$$

where "$\star$" denotes the Φ-action on $A^*(\mathfrak{a})$. Hence A^{-1} is a Φ-module isomorphism, and (49) follows.

For (50), recall that $\beta(\mathfrak{a})$ was defined with respect to some fixed element $a_0 \in \mathfrak{a}$ with $a_0 \notin (\varsigma - 1)\mathfrak{a}$. But Lemma 4.1 implies that we could use any other $b_0 \in \mathfrak{a}$ with $b_0 \notin (\varsigma - 1)\mathfrak{a}$, and we would still get a Φ-module isomorphic to $\beta(\mathfrak{a})$. For example, we could use $A^{-1}(a_0)$ to construct $\beta(A^{-1} \cdot \mathfrak{a})$ because $A^{-1}(a_0) \notin A^{-1} \cdot \mathfrak{a}$. If it were, then a_0 would be in $(\varsigma^d - 1)\mathfrak{a}$ which is $(\varsigma - 1)(1 + \varsigma + \cdots + \varsigma^{d-1})\mathfrak{a}$, and this would say that $a_0 \in (\varsigma - 1)\mathfrak{a}$ which is not true.

It therefore suffices to show that $A^*(\beta(\mathfrak{a})) \approx \beta'(A^{-1} \cdot \mathfrak{a})$ where β' is constructed using $A^{-1}(a_0)$. In the notation of Lemma 4.1 we want to show that

$$A^* \left((\mathfrak{a}, a_0)\right) \approx \left(A^{-1} \cdot \mathfrak{a}, A^{-1}(a_0)\right).$$

Recall that

$$\beta(\mathfrak{a}) = \mathfrak{a} \oplus y\mathbb{Z}$$

and

$$g \cdot y = a_0 \oplus y,$$

so

$$\beta'\left(A^{-1} \cdot \mathfrak{a}\right) = A^{-1} \cdot \mathfrak{a} \oplus y'\mathbb{Z}$$

and

$$g \cdot y' = A^{-1}(a_0) \oplus y'.$$

Define $F : \beta'(A^{-1} \cdot \mathfrak{a}) \to A^*(\beta(\mathfrak{a}))$ by $f(y) = y'$ and

$$f(x) = \left(\sum_{i=0}^{d-1} \varsigma^i\right) A(x)$$

for $x \in A^{-1} \cdot \mathfrak{a}$. We want to show that f is an isomorphism. It is clear that f is bijective, so it is enough to show that f is a Φ-homomorphism. The computation is obvious.

$$\begin{aligned} f(g \cdot x) &= f(\varsigma x) \\ &= \left(\sum_{i=0}^{d-1} \varsigma^i\right) A(\varsigma x) \\ &= \varsigma^d \left(\sum_{i=0}^{d-1} \varsigma^i\right) A(x) \\ &= g^d \cdot f(x) \\ &= g \star f(x). \end{aligned}$$

We also have

$$\begin{aligned} f(g \cdot y) &= f(A^{-1}(a_0) \oplus y) \\ &= \left(\sum_{i=0}^{d-1} \varsigma^i\right) a_0 \oplus y' \\ &= g^d \cdot y' \\ &= g \star y' \\ &= g \star f(y). \end{aligned}$$

□

Now we see that if N and N' are semi-linearly isomorphic Φ-modules, we can assume $N = N(A, B, C; [\mathfrak{a}])$ and $N' = N(A, B, C; A \cdot [\mathfrak{a}])$ for some $A \in G(= \mathrm{Aut}(\Phi))$. The case that is going to give us trouble is obviously the case $N = 1 \oplus B\mathfrak{a} = N(1, B, 0; [\mathfrak{a}])$. Φ-modules of this particular type will be called *exceptional*; the others, *non-exceptional*. We say that a Bieberbach group (with prime order holonomy) is *exceptional* if its unique maximal abelian subgroup is exceptional as a Φ-module.

We need a little more notation to state the classification theorem. Recall that G, the galois group, acts on C_p, the ideal class group. We let $\tilde{C}_p$ be the orbit set, which is not a group. Recall also that $G \approx \mathbb{Z}_{p-1}$ and $A_{1/2} \in G$ is the element with $(A_{1/2})^2 =$ the identity. Let $\tilde{C}_p^2$ be the orbit set of the action of $A_{1/2}$ on $\tilde{C}_p$.

Theorem 6.3 : *There is a one-to-one correspondence between isomorphisim classes of non-exceptional Bieberbach groups whose holonomy group has order p and 4-tuples $(A, B, C; \theta)$ where $A, B, C \in \mathbb{Z}$ with $A > 0, B \geq 0, C \geq 0, (A, C) \neq (1, 0), (B, C) \neq (0, 0)$, and $\theta \in \tilde{C}_p$.*

This theorem is an immediate consequence of the previous discussion in this chapter, but at the risk of boring the reader, we will run through it again. The Bieberbach group π satisfies

$$0 \longrightarrow M \longrightarrow \pi \longrightarrow \Phi \longrightarrow 1$$

where $\Phi \approx \mathbb{Z}_p$. By Reiner's Theorem (4.1), M is a direct sum of A modules of type 1, B of type α, and C of type β with invariants A, B, C, and $[\mathfrak{a}] \in C_p$. By Theorem 5.1, if $A > 0$, $H^2(\Phi; M) \neq 0$, so by Theorem 2.1 of Chapter III, every non-zero element in $H^2(\Phi; M)$ defines a torsionfree extension, i.e. a Bieberbach group. We must have $(B, C) \neq (0, 0)$ because M must be a faithful Φ-module, and we must have $(A, C) \neq (1, 0)$ to avoid the exceptional case.

The remaining question is which of these Bieberbach groups are isomorphic. If M and M' are semi-linearly isomorphic, then by Theorem 6.2, they have the same global invariants, and if M has local invariant $[\mathfrak{a}]$, then M' has local invariant $A \cdot [\mathfrak{a}]$ for some $A \in G$. Therefore the semi-linear isomorphism classes of Φ-modules with the same global invariants are in one-to-one correspondence with the elements of $\tilde{C}_p$. By part i) of Corollary 6.1, we can compose a semi-linear isomorphism with an isomorphism so that the composition will map any given non-zero class in $H^2(\Phi; M)$ to any non-zero class in $H^2(\Phi; M)$. Thus we can ignore the cohomology class of the extension, and Theorem 6.3 follows.

The exceptional case is not so easy.

Theorem 6.4 : *There is a one-to-one correspondence between isomorphism classes of exceptional Bieberbach groups whose holonomy group has prime order p and pairs (B, θ) where $B \in \mathbb{Z}, B > 0$, and $\theta \in \tilde{C}_p^2$.*

This is somewhat surprising since it would appear that an exceptional Bieberbach group would depend on which class in $H^2(\Phi; M)$ the extension corresponds to, and the theorem *seems* to say the contrary.

Proof: In this proof, we will confuse ideals with their ideal classes, so $\mathfrak{a}$ will be the same as $[\mathfrak{a}]$. Let $\mathfrak{a}_1, \mathfrak{a}_2, \ldots, \mathfrak{a}_q \in C_p$ be a set of orbit representatives for the action of G on C_p, i.e. if $\mathfrak{a} \in C_p$, then there is an i with $1 \leq i \leq q$ and $A \in G$ s.t. $A \cdot \mathfrak{a} = \mathfrak{a}_i$, and, furthermore, if $i \neq j$, $\mathfrak{a}_i$ is not in the same orbit as $\mathfrak{a}_j$. Fix a generator $\lambda_0 \in H^2(\Phi; M)$. Fix $B > 0$, and let S be the set of pairs $(\mathfrak{a}, \gamma)$ with $\mathfrak{a} \in C_p$ and

$$\gamma \in H^2(\Phi; N(1, B, 0; \mathfrak{a})).$$

(Since B is fixed, $\mathfrak{a}$ determines $N(1, B, 0; \mathfrak{a})$.)

We define a most uncanonical map

$$F : S \to C_p$$

as follows. Given $(\mathfrak{a}, \gamma) \in S$, there is a unique $A_1 \in G$ and i s.t.

$$A_1 \cdot \mathfrak{a} = \mathfrak{a}_i. \tag{51}$$

Let $I : \mathbb{Z} \to N(1, B, 0; \mathfrak{a}) = 1 \oplus (B-1)R \oplus \mathfrak{a}$ be the injection into the first summand. Notice that $I_* : H^2(\Phi; \mathbb{Z}) \to H^2(\Phi; N(1, B, 0; \mathfrak{a}))$ is an isomorphism. Pick $\lambda \in H^2(\Phi; \mathbb{Z})$ s.t.

$$I_*(\lambda) = \gamma. \tag{52}$$

Now pick $A_2 \in G$ s.t.

$$(A_2)_*(\lambda) = \lambda_0. \tag{53}$$

Finally put

$$F(\mathfrak{a}, \gamma) = A_1 A_2(\mathfrak{a}_i) \in C_p. \tag{54}$$

By (51), $\mathfrak{a}$ determines A_1 uniquely, and given A_1, (52) determines λ uniquely. To see that F is well-defined by (54) we must see that (53) determines A_2 uniquely. This follows from the following easy exercise:

Exercise 6.3: Suppose $A \in G$ and $A(g) = g^k$ for $g \in \Phi$. Then

$$A_*(\lambda) = k \cdot \lambda$$

for $\lambda \in H^2(\varphi; \mathbb{Z})$.

The map F is not canonical because of the arbitrary choices of the $\mathfrak{a}_i$ and λ_0.

Claim: Suppose $F(\mathfrak{a}, \gamma) = F(\mathfrak{a}', \gamma')$. Then $N(1, B, 0; \mathfrak{a})$ is semi-linearly isomorphic to $N(1, B, 0; \mathfrak{a}')$ by a semi-linear isomorphism (f, A) with

$$f_*(\gamma) = A_*(\gamma').$$

Let A_1 and A_2 (respectively A_1' and A_2') be the elements of G used in the definition of $F(\mathfrak{a}, \gamma)$ (respectively $F(\mathfrak{a}', \gamma')$), i.e. in (53) and (54). $F(\mathfrak{a}, \gamma) = F(\mathfrak{a}', \gamma')$ implies

$$A_1 A_2(\mathfrak{a}_i) = A_1' A_2'(\mathfrak{a}_j) \tag{55}$$

for some i and j. Since $\mathfrak{a}_i$ and $\mathfrak{a}_j$ are orbit representatives, we must have $i = j$, i.e. $\mathfrak{a}$ and $\mathfrak{a}'$ are in the same G-orbit. Theorem 6.2 then tells us that $N = N(1, B, 0; \mathfrak{a})$ and $N' = N(1, B, 0; \mathfrak{a}')$ are semi-linearly isomorphic.

We also know that (55) becomes

$$A_1 A_2(\mathfrak{a}_i) = A_1' A_2'(\mathfrak{a}_i) \tag{56}$$

Now let $A = A_2^{-1} A_2'$. We see that

$$\begin{aligned} A(\mathfrak{a}) &= A_2^{-1} A_2' A_1^{-1}(\mathfrak{a}_i) \\ &= (A_1')^{-1}(\mathfrak{a}_i) \\ &= \mathfrak{a}' \end{aligned} \tag{57}$$

by (56) and the definition of A_1'. In addition, we have

$$A_*(\lambda') = (A_2^{-1})_*(A_2')_*(\lambda') = (A_2^{-1})_*(\lambda_0) = \lambda.$$

Hence

$$(I')_*^{-1}[A_*(\gamma')] = A_*(\lambda) = \lambda = I_*(\gamma)$$

where $I' : \mathbb{Z} \to N'$ is the injection to the first direct summand of N'. Finally, we get

$$A_*(\gamma') = \gamma. \tag{58}$$

We know that N and N' are semi-linearly isomorphic, and we want to pick carefully a semi-linear isomorphism. We have

$$N = N(1, B, 0; \mathfrak{a})$$

and

$$N' = N(1, B, 0; \mathfrak{a}').$$

Let $f : N \to N'$ be the identity on the first two summands and equal A on the third summand, i.e. $f(n \oplus x \oplus a) = n \oplus x \oplus A(a)$ for $n \in \mathbb{Z}, x \in (B-1)R$, and $a \in \mathfrak{a}$. Since $A \cdot \mathfrak{a} = \mathfrak{a}'$ by (57), this is a map between N and N'. The fact that $A(g \cdot a) = A(g) \cdot A(a)$ shows that (f, A) is a semi-linear isomorphism between N and N'. Clearly $f_*(\gamma) = \gamma$, so by (58), we get

$$f_*(\gamma) = A_*(\gamma')$$

as desired.

Claim: Let (f, A) be a semi-linear isomorphism from $N = N(1, B, 0; \mathfrak{a})$ to $N' = N(1, B, 0; \mathfrak{a}')$ s.t. $f_*(\gamma) = A_*(\gamma')$ for some $\gamma \in H^2(\Phi; N)$ and $\gamma' \in H^2(\Phi; N')$. Then either

$$F(\mathfrak{a}, \gamma) = F(\mathfrak{a}', \gamma')$$

or

$$F(\mathfrak{a}, \gamma) = A_{1/2} \cdot F(\mathfrak{a}', \gamma').$$

Since N and N' are semi-linearly isomorphic, by Theorem 6.2, we can assume that $\mathfrak{a}' = B \cdot \mathfrak{a}$ for some $B \in G$. We can further assume that

$$f(n \oplus x \oplus a) = n \oplus x \oplus B(a)$$

for $n \in \mathbb{Z}, x \in (B-1)R$ (careful, there are two different "B's" in use now), and $a \in \mathfrak{a}$, and

$$f_*(\gamma) = \pm B_*(\gamma').$$

The "±" comes from the possibility that f may involve a sign change in the first coordinate (the only automorphism of $\mathbb{Z}$), and since we are assuming that f is the identity on the first coordinate, the "±" must appear somewhere else. Now since f is the identity on the first summand, we have $f_*(\gamma) = \gamma$, so we get

$$B_*(\gamma') = \pm\gamma \tag{59}$$

and we know that

$$B \cdot \mathfrak{a} = \mathfrak{a}'. \tag{60}$$

Let A_1, A_2, A'_1, and A'_2 be as above (i.e. in the definition of $F(\mathfrak{a}, \gamma)$ and $F(\mathfrak{a}', \gamma')$). Again we have $\mathfrak{a}_i = \mathfrak{a}_j$ since (60) says that $\mathfrak{a}$ and $\mathfrak{a}'$ are in the same orbit of G. Hence $A_2(\lambda) = \lambda_0 = A'_2(\lambda')$, so $A_2^{-1}A'_2(\lambda') = \lambda$. Notice that $(A_{1/2})_*(\lambda) = -\lambda$. It is then easily seen from this and (59) that we must have either $B = A_2^{-1}A'_2$ or $B = A_{1/2}A_2^{-1}A'_2$. By (60), we get the following string of equations:

$$\begin{aligned} A_2^{-1}A'_2(\mathfrak{a}) = \mathfrak{a}' \quad &\text{or} \quad A_{1/2}A_2^{-1}A'_2(\mathfrak{a}) = \mathfrak{a}', \\ A'_2A_1^{-1}(\mathfrak{a}_i) = A_2(A'_1)^{-1}(\mathfrak{a}_i) \quad &\text{or} \quad A_{1/2}A'_2A_1^{-1}(\mathfrak{a}_i) = A_2(A'_1)^{-1}(\mathfrak{a}_i), \\ A'_1A'_2(\mathfrak{a}_i) = A_1A_2(\mathfrak{a}_i) \quad &\text{or} \quad A_{1/2}A'_1A'_2(\mathfrak{a}_i) = A_1A_2(\mathfrak{a}_i), \\ F(\mathfrak{a}', \gamma') = F(\mathfrak{a}, \gamma) \quad &\text{or} \quad A_{1/2}F(\mathfrak{a}', \gamma') = F(\mathfrak{a}, \gamma), \end{aligned}$$

where the last equation follows from (55). Thus the claim is proved.

To finish the theorem it suffices to show that for any $\mathfrak{a} \in C_p$ we can find $(\mathfrak{a}', \gamma') \in S$ s.t. $F(\mathfrak{a}', \gamma') = \mathfrak{a}$. Let $A_1 \in G$ satisfy $A_1 \cdot \mathfrak{a} = \mathfrak{a}_i$, let

$$\rho = I_*(A_1)_*(\lambda_0) \in H^2(\Phi; 1 \oplus (B-1)R \oplus \mathfrak{a}), \tag{61}$$

and let $(\mathfrak{a}', \gamma') = (\mathfrak{a}, \rho)$. Now (61) says $A_1(\lambda_0) = \lambda$, so $A_2 = A_1^{-1}$, i.e. $A_2(\lambda) = \lambda_0$ and

$$F(\mathfrak{a}, \rho) = A_1A_2(\mathfrak{a})\mathfrak{a},$$

and we are done. □

Remark: Suppose π and π' are two non-exceptional Bieberbach groups whose holonomy has prime order. Then Theorem 6.3 says that if their subgroups of pure translation are isomorphic as Φ-modules, then π and π' are isomorphic. Theorem 6.4 seems to say that the same is true for the exceptional case, but an examination of the proof shows that this is not the case. If we think of an exceptional Bieberbach group as a triple $(B, \mathfrak{a}, \gamma)$ with $B \in \mathbb{Z}$, $\mathfrak{a}$ an ideal class in C_p, and $\gamma \in H^2(\Phi; 1 \oplus (B-1)R \oplus \mathfrak{a})$, then two such have isomorphic pure translation modules iff $B = B'$ and $\mathfrak{a} = \mathfrak{a}'$, while the corresponding Bieberbach groups are isomorphic iff $B = B'$ and there is $A \in G$ with $A \cdot \mathfrak{a} = \mathfrak{a}'$ and $A_*(\gamma') = \pm\gamma$. So, in general, the exceptional Bieberbach group $(B, \mathfrak{a}, \gamma)$ is not isomorphic to $(B, \mathfrak{a}, \gamma')$ even though the action of Φ on the pure translations is the "same."

7. $\mathbb{Z}_p$-manifolds

The classification theorems of the previous section have immediate consequences for flat manifolds. The following notation is extremely convenient. For this definition we let Φ be any group, finite or infinite, discrete or continuous.

Definition 7.1: A Φ-*manifold* is a compact riemannian manifold whose holonomy group is isomorphic to Φ.

One could leave out the condition of compactness to get a somewhat more general notion. If Φ is finite or even totally disconnected, a Φ-manifold will be flat. The theorems of the previous section yield classification theorems for $\mathbb{Z}_p$-manifolds up to affine equivalence. (Recall that an affine equivalence is a diffeomorphism that preserves the riemannian connection.) We collect these results here.

Theorem 7.1 : *There is a one-to-one correspondence between affine equivalence classes of* $\mathbb{Z}_p$*-manifolds and 4-tuples* $(A, B, C; \theta)$ *with* $A, B, C \in \mathbb{Z}$, $A > 0, B \geq 0, C \geq 0, BC \neq 0$, *and* $\theta \in \tilde{C}_p$ *if* $(A, C) \neq (1,0)$ *or* $\theta \in \tilde{C}_p^2$ *if* $(A, C) = (1,0)$.

We use $X(A, B, C; \theta)$ to denote the (affine equivalemce class of a) $\mathbb{Z}_p$-manifold with invariants $(A, B, C; \theta)$. The invariants A, B, and C are easy to see geometrically.

Exercise 7.1: i) Show that the dimension of $X(A, B, C; \theta)$ is $A+(p-1)B + pC$.

ii) If π is any group show that $H_1(\pi; \mathbb{Z}) \approx \pi/[\pi, \pi]$.

iii) If

$$0 \longrightarrow M \longrightarrow \pi \longrightarrow \Phi \longrightarrow 1$$

defines a Bieberbach group with Φ cyclic, show

$$\frac{\pi}{[\pi,\pi]} \approx \frac{N}{\Delta n}.$$

Is this true if Φ is not cyclic?

iv) Show that

$$H_1(X(A,B,C;\theta);\mathbb{Z}) \approx \mathbb{Z}^{A+C} \oplus \mathbb{Z}_p^B$$

and

$$H^1(X(A,B,C;\theta);\mathbb{Z}) \approx \mathbb{Z}^{A+C}.$$

v) If $h_p = 1$ (i.e. C_p is the trivial group), show that the number of (affine equivalence classes of) n-dimensional $\mathbb{Z}_p$-manifolds is

$$\frac{1}{2}\left(\left[\frac{n-1}{p}\right]^2 + 3\left[\frac{n-1}{p}\right]\right) +$$
$$\frac{1}{2}\left\{\left(\left[\frac{n-1}{p-1}\right] - \left[\frac{n-1}{p}\right]\right)\left(2n-(p-1)\left(\left[\frac{n-1}{p-1}\right]+\left[\frac{n-1}{p}\right]+1\right)\right)\right\}.$$

If $h_p \neq 1$, the number of $\mathbb{Z}_p$-manifolds in dimension n will depend on the size of $\tilde{C}_p$ and $\tilde{C}_p^2$.

vi) Show that $X(A,B,C;\theta)$ is orientable iff

$$\left(\frac{p-1}{2}\right)(B+C) \equiv 0 \text{ (mod2)}$$

for $p \neq 2$. If $p = 2$ the condition is $(B+C) \equiv 0$ (mod2).

Remark: Parts i) and iv) of the above exercise show that if we know the dimension of a $\mathbb{Z}_p$-manifold X and $H_1(X;\mathbb{Z})$, then we know A, B, C because the equations

$$n = A + (p-1)B + pC,$$
$$\beta_1 = A + C,$$

and

$$\tau_1^p = B$$

can be solved for A, B, and C where n is the dimension of X, β_1 is the first Betti number, and τ_1^p is the dimension over $\mathbb{Z}_p$ of the p-torsion of

$H_1(X;\mathbb{Z})$. In fact, the entire cohomology of X can be computed (at least additively), and depends on only the global invariants A, B, and C. See [19] for a discussion of this result. At the present writing no geometric interpretation of the local invariant θ is known. $\mathbb{Z}_p$-manifolds with the same global invariants are remarkedly similar. A striking example of this phenomenon follows in the next section.

8. An Interesting Example

The only flat manifold in dimension one is, of course, the circle. It is, after all, the only compact manifold there. The circle S^1 corresponds to the Bieberbach group $\mathbb{Z}$. Let $X = X(A,B,C;\theta)$ be a $\mathbb{Z}_p$-manifold with the notation as in the previous section. Then

$$\pi_1(X \times S^1) \approx \pi_1(X) \times \mathbb{Z},$$

so $X \times S^1 = X(A+1,B,C;\theta)$ where (and this is the important point here) if $(A,C) = (1,0)$, we must project θ from $\tilde{C}_p^2$ to $\tilde{C}_p$ to get the local invariant for $X \times S^1$. Thus $X \times S^1$ is never exceptional even if X is.

For example, let's take $A = 1, B = 1$, and $C = 0$. Thus we are in the lowest possible dimension in which a $\mathbb{Z}_p$-manifold can exists, namely p. Now suppose we have ideal classes $[\mathfrak{a}]$ and $[\mathfrak{a}']$ in C_p with the property that $A_{1/2} \cdot [\mathfrak{a}] \neq [\mathfrak{a}']$, but that there is $A \in G$ with $A \cdot [\mathfrak{a}] = [\mathfrak{a}']$. Then $X = X(1,0,0;[\mathfrak{a}])$ would not be the same as (i.e. affinely equivalent to) $X' = X(1,0,0;[\mathfrak{a}'])$ since $[\mathfrak{a}]$ and $[\mathfrak{a}']$ determine different elements of $\tilde{C}_p^2$. In fact, X and X' are of different homotopy types since their fundamental groups are not isomorphic. On the other hand, $X \times S^1$ and $X' \times S^1$ are not only homeomorphic, but actually affine equivalent since $[\mathfrak{a}]$ and $[\mathfrak{a}']$ do determine the same class in $\tilde{C}_p$.

Therefore, to get our interesting example, we need ideals (or ideal classes) with the right properties. To see that these exist requires the introduction of some new material which will turn out to be useful in the next chapter. We will, therefore, do some of this material in more detail than required here, although we will only give references for some of the proofs, some of which are quite deep.

First of all it is clear that $A_{1/2} \cdot \mathfrak{a} = \bar{\mathfrak{a}}$, but we may have $[\mathfrak{a}] = [\bar{\mathfrak{a}}]$ without having $\mathfrak{a} = \bar{\mathfrak{a}}$, i.e. $[\mathfrak{a}] = [\bar{\mathfrak{a}}]$ means that $\mathfrak{a}$ and $\bar{\mathfrak{a}}$ are isomorphic as $\mathbb{Z}_p$-modules, while $\mathfrak{a} = \bar{\mathfrak{a}}$ means that they are the *same* $\mathbb{Z}_p$-module, i.e. the same set. Although the notation is by no means standardized, we make the following definitions:

Definition 8.1: If $\mathfrak{a} = \bar{\mathfrak{a}}$, we say $\mathfrak{a}$ is an *ambiguous* ideal. If $[\mathfrak{a}] = [\bar{\mathfrak{a}}]$, we say that $[\mathfrak{a}]$ is an *ambiguous* ideal class. If $[\mathfrak{a}]$ contains an ambiguous ideal, we say that $[\mathfrak{a}]$ is a *strongly ambiguous* ideal class.

Clearly a strongly ambiguous ideal class is ambiguous. We let R_0 be the subring $\mathbb{Z}[\varsigma + \varsigma^{-1}]$ of $R = \mathbb{Z}[\varsigma]$.

Exercise 8.1: Show that R_0 is the maximal real subring of R.

Suppose I is an ideal in R_0. Then $R \cdot I$ is some ideal of R. We say that $\mathfrak{a}$, an ideal in R, *comes from* R_0 if $\mathfrak{a} = R \cdot I$ for some ideal I in R_0.

Proposition 8.1: *An ideal class* $[\mathfrak{a}] \in C_p$ *is strongly ambiguous iff* $[\mathfrak{a}]$ *contains an ideal which comes from* R_0.

Proof: Clearly an ideal which comes from R_0 is ambiguous, so its class is strongly ambiguous. Thus it only remains to show that an ambiguous ideal comes from R_0.

Suppose $\mathfrak{a} = \bar{\mathfrak{a}}$. By the Fundamental Theorem of Arithmetic for R (Theorem 2.3), we can write

$$\mathfrak{a} = \mathfrak{p}_1^{e_1}\mathfrak{p}_2^{e_2}\cdots\mathfrak{p}_k^{e_k} \tag{63}$$

where $\mathfrak{p}_1, \ldots, \mathfrak{p}_k$ are the distinct prime ideals that divide $\mathfrak{a}$. Suppose $\mathfrak{p} = (\varsigma - 1)$ is one of the $\mathfrak{p}_i$'s. We have

$$\bar{\mathfrak{p}} = (\bar{\varsigma} - 1) = (\varsigma^{-1} - 1) = (-\varsigma^{-1}(\varsigma - 1)) = (\varsigma - 1) = \mathfrak{p},$$

so $\mathfrak{p}$ is ambiguous, and we can divide $\mathfrak{a}$ by a suitable power of $\mathfrak{p}$ getting an ideal in the same ideal class as $\mathfrak{a}$ which is still ambiguous. Hence we may as well assume that $\mathfrak{p}$ does not divide $\mathfrak{a}$.

Since $\mathfrak{a} = \bar{\mathfrak{a}}$, we can arrange the notation in (63) so that

$$\mathfrak{a} = \mathfrak{p}_1^{e_1}\bar{\mathfrak{p}}_1^{e_1}\cdots\mathfrak{p}_l^{e_l}\bar{\mathfrak{p}}_l^{e_l}\mathfrak{q}_1^{f_1}\cdots\mathfrak{q}_m^{f_m} \tag{64}$$

where $\bar{\mathfrak{p}}_i \neq \mathfrak{p}_i$ and $\bar{\mathfrak{q}}_i = \mathfrak{q}_i$. Let $P_i = \mathfrak{p}_i \cap R_0$ and $Q_i = \mathfrak{q}_i \cap R_0$.

Claim: $P_iR = \mathfrak{p}_i\bar{\mathfrak{p}}_i$ and $Q_iR = \mathfrak{q}_i$.

The P_i's and the Q_i's are prime ideals of R_0. We are asking what the prime decomposition of P_i and Q_i are when lifted to R. Let P be any prime ideal of R_0 and write

$$PR = \mathfrak{b}_1^{g_1} \cdots \mathfrak{b}_n^{g_n}$$

where $\mathfrak{b}_i$ is prime in R. We say P *ramifies* in R if any g_i is greater than 1.

Lemma 8.1: *If a prime ideal P of R_0 ramifies in R, then $(\varsigma - 1)$ divides PR.*

Exercise 8.2: Assuming the lemma, prove the claim. (**Hint:** Recall we have arranged not to have $(\varsigma - 1)$ as a factor of $\mathfrak{a}$.)

Now the proposition follows easily since (64) can be rewritten as

$$\mathfrak{a} = (P_1 \cdots P_l Q_1 \cdots Q_m)R,$$

which shows that $\mathfrak{a}$ comes from R. □

To prove Lemma 8.1 we need a tool which can tell us about ramification. Let A be any subring of R (like R_0) s.t. R is a free module of rank n over A.

Definition 8.2: Let $r \in R$. Recall that the trace of r, $\mathrm{tr}_{R/A}(r)$, is the trace of the endomorphism obtained by multiplication by r considering R as a free module over A. Let $r_1, \ldots, r_n \in R$. The *discriminant* of $r_1, \ldots, r_n$ (with respect to A) is the element of A given by

$$D(r_1, \ldots, r_n) = \det\left(\mathrm{tr}_{R/A}(r_i r_j)\right).$$

Let's do an appropriate example. Take $A = R_0$, so R is a free module of rank 2 over A. It is easy to see that $\{1, \varsigma\}$ is a basis for R over A. Let's compute $D(1, \varsigma)$. $\mathrm{Tr}_{R/A}(1) = 2$ while $\mathrm{tr}_{R/A}(\varsigma) = \varsigma + \varsigma^{-1}$ as can be seen from Proposition 2.2 or computed directly, i.e. multiplication by ς sends 1 to ς and ς to $\varsigma^2 = (\varsigma + \varsigma^{-1})\varsigma - 1$. So with respect to the basis $\{1, \varsigma\}$, multiplication by ς has the matrix

$$\begin{pmatrix} 0 & -1 \\ 1 & \varsigma + \varsigma^{-1} \end{pmatrix}.$$

We see that $\mathrm{tr}_{R/A}(\varsigma) = \varsigma + \varsigma^{-1}$. Similarly $\mathrm{tr}_{R/A}(\varsigma^2) = \varsigma^2 - \varsigma^{-2}$. Thus

$$D(1,\varsigma) = \det\begin{pmatrix} 1 & \varsigma+\varsigma^{-1} \\ \varsigma+\varsigma^{-1} & \varsigma^2+\varsigma^{-2} \end{pmatrix} = (\varsigma - \varsigma^{-1})^2.$$

Notice that this is real since $\varsigma^{-1} = \bar{\varsigma}$.

Exercise 8.3: If A is any appropriate subring of R and $\{r_1, \ldots, r_n\}$ is any basis of R over A, then show that the principal ideal of A generated by $D(r_1, \ldots, r_n)$ is independent of the choice of the basis $\{r_1, \ldots, r_n\}$. (**Hint:** This is straight linear algebra.)

Definition 8.3: We call the ideal generated by $D(r_1, \ldots, r_n)$ where $\{r_1, \ldots, r_n\}$ is any basis of R over A the *discriminant* of R over A and denote it by $\mathcal{D}_{R/A}$.

The main fact we need about discriminants is the following:

Theorem 8.1 : *If A is any subring of R s.t. R is a free module over A, then a prime ideal P of A ramifies in R iff the discriminant ideal $\mathcal{D}_{R/A}$ divides P.*

A proof of this result can be found on page 74 of [72]. The idea is to divide everything by P and work with the finite fields R/PR and A/P. This omitted proof is not one of the deep ones.

Now we can give the proof of Lemma 8.1.

Proof: We can see from the example worked out above that

$$\mathcal{D}_{R/R_0} = \left((\varsigma - \varsigma^{-1})^2\right).$$

□

Now here is a way (due to Dick Gross) to get the ideal class we want. Take a non-trivial ambiguous ideal class $[\mathfrak{a}]$ and look at $A \cdot [\mathfrak{a}]$ for each $A \in G$, the galois group. If $A \cdot [\mathfrak{a}] = [\mathfrak{a}]$ for all A, then we can show that $[\mathfrak{a}]$ must contain an ideal which comes from $\mathbb{Z}$, but all ideals of $\mathbb{Z}$ are principal. Since $[\mathfrak{a}]$ is non-trivial, there must be at least one A s.t. $A \cdot [\mathfrak{a}] \neq [\mathfrak{a}]$. Since $[\mathfrak{a}]$ is ambiguous, $A_{1/2} \cdot [\mathfrak{a}] = [\mathfrak{a}]$, so $A_{1/2} \cdot [\mathfrak{a}] \neq A \cdot [\mathfrak{a}]$, and we have found ideal classes, namely $[\mathfrak{a}]$ and $A \cdot [\mathfrak{a}]$ which have the desired properties.

To make this scheme work, we need to know the following:

i) There is a prime number p s.t. h_p (the order of C_p) is not 1.

ii) There is a prime number p s.t. C_p contains a non-trivial ambiguous ideal.
iii) If $A \cdot [\mathfrak{a}] = [\mathfrak{a}]$ for $A \in G$, then $[\mathfrak{a}]$ contains an ideal that comes from $\mathbb{Z}$, i.e. $[\mathfrak{a}] = nR$ for some $n \in \mathbb{Z}$.

There are various ways to get i) and ii); all of them use more-or-less deep results. For example, if we knew that $h_{29} = 8$ (which it does), that settles i). Since the identity element of C_p is surely ambiguous, there must be another (i.e. non-trivial) ambiguous class because there are 7 non-trivial classes, and there is no way that seven classes can all be non-ambiguous.

Actually we can avoid the difficult result $h_{29} = 8$. All we needed was the fact that h_{29} was even. It turns out that, in general, h_p is the product of two factors $h_p^{(1)}$ and $h_p^{(2)}$, i.e. $h_p = h_p^{(1)} h_p^{(2)}$. There are again various ways to see this. One can write down a formula for h_p and then factor this formula. (see [11], page 358). Alternatively, one can let $h_p^{(2)}$ equal the class number of R_0 and then show that $h_p^{(2)}$ divides h_p (cf. Theorem 8.2 below). In any case, it turns out that $h_p^{(1)}$ is much easier to compute than $h_p^{(2)}$. Kummer, in 1851 (see [51]), computed $h_p^{(1)}$ for all primes less than 100, and by 1964, $h_p^{(1)}$ was known for $p \leq 257$ (see [73]). However, as of 1968, the only information known about $h_p^{(2)}$ was that it was 1 for $p \leq 19$. The point is that Kummer showed that $h_{29}^{(1)} = 8$, so h_{29} must be even. Thus we can use the above argument without knowing anything at all about $h_{29}^{(2)}$.

Another way to get i) and ii) is to consider the norm map from R down to R_0, i.e. if $[\mathfrak{a}] \in C_p$, then we can consider $[\mathfrak{a}\bar{\mathfrak{a}}] \in \Gamma_p$, the ideal class group of R_0.

Exercise 8.4: Show that the map $N : C_p \to \Gamma_p$ defined by $N([\mathfrak{a}]) = [\mathfrak{a}\bar{\mathfrak{a}}]$ is a well-defined homomorphism.

We also have the following:

Theorem 8.2 : *N is surjective.*

We will give references later.

If we take any non-principal ideal P in R_0 and lift it to R, we will get an ambiguous ideal $\mathfrak{a}$ in R whose class is ambiguous, in fact, strongly ambiguous. Then we would need to know that there is a prime p s.t. Γ_p is not trivial. This is true, but hard. One approach is to observe that R_0 contains a quadratic ring Q, i.e. a subring of rank 2 over $\mathbb{Z}$. Quadratic rings are much easier to work with, and i) and ii) can be handled thusly.

We are left with iii). We state it as a proposition.

Proposition 8.2: *Suppose $[\mathfrak{a}] \in C_p$ satisfies*

$$A \cdot [\mathfrak{a}] = [\mathfrak{a}]$$

for all $A \in G$, the galois group of R over $\mathbb{Z}$. Then there is an ideal $\mathfrak{a} \in [\mathfrak{a}]$ s.t.

$$\mathfrak{a} = nR$$

for some $n \in \mathbb{Z}$.

We are going to sketch the proof of this result (which is not deep), leaving parts as an exercise. We have stated and proved Proposition 8.1 not only because it will be needed in Chapter V, but also because its proof can be copied for Proposition 8.2. After all, we can consider the galois theory of R over R_0. The galois group of R over R_0 has just two elements, 1 and $A_{1/2}$. Since $A_{1/2} \cdot [\mathfrak{a}] = [\bar{\mathfrak{a}}]$, Propositions 8.1 and 8.2 say pretty much the same thing.

We can again write

$$\mathfrak{a} = \mathfrak{p}_1^{e_1} \mathfrak{p}_2^{e_2} \cdots \mathfrak{p}_k^{e_k}. \tag{65}$$

If $A \in G$, then $A \cdot \varsigma = \varsigma^c$ for some c with $1 \le c \le p-1$, so

$$A \cdot (\varsigma - 1) = (\varsigma^c - 1) = \left(\frac{\varsigma^c - 1}{\varsigma - 1}\right)(\varsigma - 1) = (\varsigma - 1)$$

because by Exercise 4.8, $(\varsigma^c - 1)/(\varsigma - 1)$ is a unit in R so

$$\left(\frac{\varsigma^c - 1}{\varsigma - 1}\right) = R.$$

Hence we can assume that the ideal $(\varsigma - 1)$ does not appear in the factorization (65).

The difficulty in the proof is mainly notational. When we try to write down the analogue of (64), things get rather complicated if not totally out of hand. We are going to leave that to you and just prove the analogue of Lemma 8.1.

Lemma 8.2: *The only prime number that ramifies in R is p.*

Proof: By Theorem 8.1, it suffices to prove

$$D(1, \varsigma, \ldots, \varsigma^{p-2}) = \pm p^{p-2}. \tag{66}$$

To do this we will use the following wonderful formula:

$$D(1, \varsigma, \ldots, \varsigma^{p-2}) = (-1)^{\frac{1}{2}(p-1)(p-2)} N(F'(\varsigma)) \tag{67}$$

where N is the usual norm (down to $\mathbb{Z}$) and F' is the derivative of the cyclotomic polynomial $X^{p-1} + X^{p-2} + \cdots + X + 1$. First we prove this wonderful formula, and then we will use it. By Proposition 2.2, we can write

$$\operatorname{tr}(\varsigma^i) = \sum_{A \in G} A \cdot \varsigma^i = \sum_{k=0}^{p-2} A_k \cdot \varsigma^i$$

where $A_k \in G$ is defined by $A_k(\varsigma) = \varsigma^k$. We have

$$\begin{aligned}
D(1, \varsigma, \ldots, \varsigma^{p-2}) &= \det\left(\operatorname{tr}(\varsigma^i \varsigma^j)\right) \\
&= \det\left(\sum_{k=0}^{p-2} A_k(\varsigma^i \varsigma^j)\right) \\
&= \det\left(\sum_{k=0}^{p-2} A_k(\varsigma^i) A_k(\varsigma^j)\right) \\
&= \left\{\det(A_k(\varsigma^i))\right\}\left\{\det(A_k(\varsigma^j))\right\} \\
&= \left\{\det(A_k(\varsigma^i))\right\}^2 \\
&= \det(\varsigma_k^j)^2 \quad \text{where we have put } \varsigma_k = A_k(\varsigma) \\
&= \left[\prod_{k<j} (\varsigma_k - \varsigma_j)\right]^2 \\
&\qquad\qquad \text{by the Vandermonde identity} \\
&= c \prod_{k \neq j} (\varsigma_k - \varsigma_j) \quad \text{where } c = (-1)^{\frac{1}{2}(p-1)(p-2)} \\
&= c \prod_{k=0}^{p-2} \left[\prod_{\substack{j=0 \\ j \neq k}}^{p-2} (\varsigma_k - \varsigma_j)\right]
\end{aligned}$$

$$= c\prod_{k=0}^{p-2} F'(\varsigma_k) \quad \text{since } F(X) = \textstyle\prod_{i=0}^{p-2}(X - \varsigma_i)$$

$$= N(F'(\varsigma)) \qquad \text{by Proposition 2.2.}$$

To use (67) we recall that

$$(X-1)F(X) = X^p - 1,$$

so

$$F(X) + (X-1)F'(X) = pX^{p-1},$$

and

$$(\varsigma - 1)F'(\varsigma) = p\varsigma^{p-1}. \tag{68}$$

By Exercise 2.2, $N(p) = p^{p-1}$, and $N(\varsigma) = \pm 1$, and by Exercise 2.4, $N(\varsigma - 1) = \pm p$. Thus we get

$$\pm pN(F'(\varsigma)) = \pm p^{p-1}(\pm 1)^{p-1},$$

so

$$pN(F'(\varsigma)) = \pm p^{p-2},$$

and (66) and the proposition follows. □

Since in R

$$p = (1-\varsigma)(1-\varsigma^2)\cdots(1-\varsigma^{p-1})$$

by Equation (7), and since we have arranged that $(\varsigma - 1)$ does not appear in our factorization (65), we don't have to worry about ramification.

Exercise 8.5: By copying the proof of Proposition 8.1 and using Lemma 8.2, prove Proposition 8.2.

We now state our result on the interesting example in a theorem. We will conclude this section with a further discussion of the number theory and by giving some references.

Theorem 8.3 : *There is a prime p (29 will do) and two $\mathbb{Z}_p$-manifolds X and X' of dimension p s.t. X and X' have non-isomorphic fundamental groups π and π' respectively (and hence X and X' are of different homotopy types), but such that $X \times S^1$ and $X' \times S^1$ are affinely equivalent (and hence homeomorphic), and consequently, $\pi \times \mathbb{Z}$ is isomorphic to $\pi' \times \mathbb{Z}$.*

Remarks: i) The class number h_p turns up in several different branches of mathematics. It is somewhat surprising to see it here in what is essentially a problem in differential geometry. It also arises in several places in differential topology (see [35] and [62]). It is perhaps less surprising to see it appear in algebraic K-theory. In fact, some of the best collections of results on h_p are to found in Milnor's work on algebraic K-theory (see [62], page 31, and [63], page 413).

What about the results we used in the proof of Theorem 8.3? The factorization $h_p = h_p^{(1)} h_p^{(2)}$ was proved by Kummer in 1851 when he also computed $h_p^{(1)}$ for $p < 100$. Since all we really needed was some even $h_p^{(1)}$, that's enough for us. A more readable version of the theory is given in Section 5 of Chapter V of [11]. One can see that $h_{29}^{(1)} = 8$ from their table 9 although it does not appear that they actually do the computation. At the time of the writing of this book (1986), the best results on $h_p^{(1)}$ was [55] which computed it for $p < 521$.

As for $h_p^{(2)}$, the best results we know of is due to van der Linden [56], namely that $h_p^{(2)} = 1$ for $p < 68$. For the result that $h_p^{(2)} \neq 1$ for some p, the authors of [2] use the idea involving the quadratic subring, while results using cubic subfields can be found in [37] and [75], and [27] and [74] use both. Incidently, [74] shows that $h_p^{(2)}$ can be larger than p which wasn't known before. The p which they use is 1129001877. If the generalized Riemann hypothesis is assumed, then it can be shown that $h_p^{(2)} = 1$ for $p < 163$, and that $h_{163}^{(2)} = 4$. Kummer never did compute a non-trivial $h_p^{(2)}$, but in 1876, he showed that $h_{163}^{(2)}$ is even. It has been conjectured that $h_p^{(2)} = 1$ for $p < 97$, but $h_{97}^{(2)} \neq 1$ (see [73], page 4).

It should be remarked that we are pretty lucky that $h_{29}^{(1)}$ is even. Schrutka von Rechtenstumm showed in [73] that for $p \leq 257$, only $h_{29}^{(1)}$, $h_{113}^{(1)}, h_{163}^{(1)}, h_{197}^{(1)}$, and $h_{239}^{(1)}$ are even. By the way, $h_{239}^{(1)}$ is approxmately 2×10^{49}.

Why is there so much interest in h_p? Well, besides the relation of h_p to the failure of unique factorization in R, we have the following famous conjecture of Fermat·

Conjecture (Fermat): There are no integers $a, b, c > 0$ and $n > 2$ s.t.

$$a^n + b^n = c^n.$$

In 1851, Kummer showed that if $p \nmid h_p$, then there are no solutions of $a^p + b^p = c^p$. There are various other connections between this conjecture and h_p and, in fact, C_p. For Fermat's conjecture see [30] and [70]. Some very important work on this problem has been done since the publication of these two books. It is due to Faltings ([31]) and the application to the Fermat conjecture can be found in [42]. For cyclotomic fields, see [53], [54], [41], and especially [77].

ii) Conner and Raymond have studied spaces X and Y which are "different," but with $X \times S^1$ the "same" as $Y \times S^1$ (see [25]). Their theory vastly generalizes Theorem 8.3.

In addition, Hilton, Mislin, and Roitberg (see [43]) have shown that there are "different" spaces X and Y s.t. $X \times S^3$ is the "same" as $Y \times S^3$. This phenomenon clearly does not depend solely on the fundamental group.

9. The Riemannian Structure of Some $\mathbb{Z}_p$-manifolds

The $\mathbb{Z}_p$-manifolds to be studied here are the ones that appeared in the previous section, namely the ones of lowest possible dimension. They correspond to the Φ-module $1 \oplus \mathfrak{a}$ where $\mathfrak{a}$ is some ideal in $R = \mathbb{Z}[\varsigma]$. If $p = 2$, there is only one such, and it is the usual two-dimensional Klein bottle. This suggests the following definition:

Definition 9.1: Let $\mathfrak{a}$ be an ideal in R. The *generalized Klein bottle* of $\mathfrak{a}$ is the differential p-manifold $K(\mathfrak{a})$ which corresponds to the class of $\mathfrak{a}$ in $\tilde{C}_p^2$. $K(\mathfrak{a})$ has a flat riemannian structure with holonomy group of order p.

We abbreviate "generalized Klein bottle" as GKB and call $K(\mathfrak{a})$ the *standard* GKB if $\mathfrak{a}$ is principal. If $\mathfrak{a}$ is not principal, we say $K(\mathfrak{a})$ is an *exotic* GKB.

We know from Theorem 7.1 that $K(\mathfrak{a})$ is diffeomorphic to $K(\mathfrak{a}')$ iff $[\mathfrak{a}] = [\mathfrak{a}']$ or $[\bar{\mathfrak{a}}] = [\mathfrak{a}']$. We want to think of $K(\mathfrak{a})$ as a differential (not riemannian) manifold because we want to study the collection of flat riemannian structures that $K(\mathfrak{a})$ supports. To do this, we recall how $K(\mathfrak{a})$ is constructed from $\mathfrak{a}$.

First, let's take $\mathfrak{a} = R$. We can define an embedding $F : R \to \mathbb{R}^{p-1}$ by setting $F(\varsigma^i) = e_{i+1}$ for $i = 0, 1, \ldots, p-2$. Since R is free as an abelian group on the basis $\{1, \varsigma, \ldots, \varsigma^{p-2}\}$, we can define F on the rest of $\mathbb{R}$ by linearity. Hence $F(R) = L^{p-1}$, the lattice of integral vectors (i.e. vectors with integer components) in $\mathbb{R}^{p-1}$. Consider $\mathbb{R}^{p-1}$ embedded in $\mathbb{R}^p$ as the first $p-1$ dimensions, i.e. write $\mathbb{R}^p = \mathbb{R}^{p-1} \oplus \mathbb{R}$. Define an affine motion $M \in \mathcal{A}_p$ by the following: On L^{p-1}, M is multiplication by ς. We extend M to R^{p-1} by linearity. On the last coordinate M is translation by $1/p$, i.e.

$$M \cdot (x \oplus z) = \varsigma \cdot x \oplus (z + \frac{1}{p})$$

for $x \in \mathbb{R}^{p-1}$ and $z \in \mathbb{R}$. By "ς" we mean here the map with matrix

$$A = \begin{pmatrix} 0 & 0 & 0 & \cdots & 0 & -1 \\ 1 & 0 & 0 & \cdots & 0 & -1 \\ 0 & 1 & 0 & \cdots & 0 & -1 \\ 0 & 0 & 1 & \cdots & 0 & -1 \\ \vdots & & & & & \vdots \\ 0 & 0 & 0 & \cdots & 1 & -1 \end{pmatrix}.$$

(Recall that $\varsigma^{p-1} = -1 - \varsigma - \cdots - \varsigma^{p-2}$.) We see that $M \in \mathcal{A}_p$. Let $\pi(R)$ be the subgroup of $\mathcal{A}_p$ generated by M together with the translations $x \oplus z \mapsto (x + e_i) \oplus z$ for $i = 1, \ldots, p-1$.

Exercise 9.1: Show that $\pi(R)$ is a Bieberbach group. In fact, you can show that $\pi(R)$ is the Bieberbach group corresponding to the $\mathbb{Z}_p$-manifold $1 \oplus R$.

The situation is a bit confusing since M was an affine transformation and not, as we would have expected, a rigid motion. However, $r(\pi(R))$ is finite; in fact, it is $\mathbb{Z}_p$. Thus we can easily change coordinates so that $\pi(R)$ is conjugated into $\mathcal{M}_p$, the group of rigid motions. Alternatively, we can change the inner product on $\mathbb{R}^p$ so that the new one is left invariant by M. Then $\pi(R)$ will be in the group of "rigid motions" of the new inner product. We prefer this approach because it enables us to construct easily all the flat riemannian structures on $\mathbb{R}^p/\pi(R)$.

Before we do that, however, we still must see how the exotic GKB's are constructed. Let $\mathfrak{a}$ be an ideal in R.

Exercise 9.2: Show that $\mathfrak{a}$ is a free abelian group of rank p. **(Hint:** Look at $R/\mathfrak{a}$.)

Although we may not be able to put our hands on a $\mathbb{Z}$-basis for $\mathfrak{a}$, there is one, say $\{a_1, \ldots, a_{p-1}\}$. We can define an embedding $F : \mathfrak{a} \to \mathbb{R}^{p-1}$ by $F(a_i) = e_i$ for $i = 1, \ldots, p-1$, and proceed exactly as in the previous case in which $\mathfrak{a}$ was R. Denote the Bieberbach group obtained in this manner by $\pi(\mathfrak{a})$. It is easy to see that $K(\mathfrak{a}) = \mathbb{R}^p/\pi(\mathfrak{a})$.

Theorem 9.1 : *For $p > 2$, the space of flat riemannian structures on $K(\mathfrak{a})$ is homeomorphic to the space $(\mathbb{R}^+)^{\frac{p+1}{2}}$ where $\mathbb{R}^+ = \{x \in \mathbb{R} : x > 0\}$.*

Proof: We must show that the space of flat riemannian structrures on $K(\mathfrak{a})$ is an open convex cone in $\mathbb{R}^{\frac{p+1}{2}}$. The discussion above shows that if suffices to find all M-invariant metrics on $\mathbb{R}^p$. It is equivalent to find all ς-invariant positive definite symmetric bilinear forms on $\mathfrak{a}$ because if we do that the only remaining variable will be the length of the translation in the last component (which we had taken to be $1/p$). This translation can be any non-zero number. Since translation by $-z$ gives the same orbit space as translation by z, this yields one of the factors $\mathbb{R}^+$. The other factors are provided by the following lemma due to Han Sah:

Lemma 9.1: *Let $\mathfrak{a}$ be an ideal in R. Then the space of positive definite symmetric bilinear forms on $\mathfrak{a}$ which are invariant under multiplication by ς is an open convex cone in $\mathbb{R}^{\frac{p+1}{2}}$.*

This concludes the proof of the theorem. $\square$

We give a sketch of the proof of the lemma. Let B be the space of all bilinear forms on $\mathfrak{a}$ with values in R, so

$$B \approx \mathrm{Hom}_{\mathbb{Z}}(\mathfrak{a} \otimes_{\mathbb{Z}} \mathfrak{a}, \mathbb{R})$$

which is a module over the ring $\mathbb{R}[\mathbb{Z}_p]$, where a generator of $\mathbb{Z}_p$ acts on $\mathfrak{a}$ by multiplication by ς. The trick is to use tensor product identities to get everything to be an $\mathbb{R}$-module. Hence we replace $\mathfrak{a}$ by $\mathfrak{a} \otimes_{\mathbb{Z}} \mathbb{R}$ to get

$$B \approx \mathrm{Hom}_{\mathbb{R}}((\mathfrak{a} \otimes_{\mathbb{Z}} \mathbb{R}) \otimes_{\mathbb{R}} (\mathfrak{a} \otimes_{\mathbb{Z}} \mathbb{R}), \mathbb{R}).$$

Let's look at $(\mathfrak{a} \otimes_{\mathbb{Z}} \mathbb{R})$ as a module over $\mathbb{R}[\mathbb{Z}_p]$. At this point it matters not what the ideal $\mathfrak{a}$ is; we could just as well take $\mathfrak{a} = R$. By putting the matrix A defined above into orthogonal normal form we see that

$$(\mathfrak{a} \otimes_{\mathbb{Z}} \mathbb{R}) \approx \bigotimes_{i=1}^{\frac{p-1}{2}} N_i$$

where each N_i is a 2-dimensional (over $\mathbb{R}$) irreducible module over $\mathbb{R}[\mathbb{Z}_p]$.

Next we observe that

$$B^{\mathbb{Z}_p} \approx \bigoplus_{i=1}^{\frac{p-1}{2}} (N_i \otimes_{\mathbb{R}} N_i)^{\mathbb{Z}_p}.$$

We have omitted the terms of the form $N_i \otimes_{\mathbb{R}} N_j$ for $i \neq j$ because they do not have any invariant elements. One can see this by looking at the representation over $\mathbb{C}$ where it breaks up into 1-dimensional representations generated by $\varsigma^{\pm i \pm j}$ for $1 \leq i \neq j \leq (p-1)/2$. These 1-dimensional complex representations have no invariant elements.

Now we look for the symmetric forms in $N_i \otimes_{\mathbb{R}} N_i$. $\bigwedge_{\mathbb{R}}^2 N_i$ is, of course, $\mathbb{R}$, and this is the subspace of skew-symmetric forms which are invariant. Again by passing to $\mathbb{C}$, we can see that the space of invariant forms in $N_i \otimes_{\mathbb{R}} N_i$ has real dimension 2, so the space of symmetric invariant forms in each $N_i \otimes_{\mathbb{R}} N_i$ has dimension 1. (It is generated by the usual one.)

Therefore the positive definite symmetric invariant forms in $N_i \otimes_{\mathbb{R}} N_i$ form a ray. When we take the direct sum over $1 \leq i \leq (p-1)/2$, we get an open convex cone in $\mathbb{R}^{\frac{p-1}{2}}$. This concludes the sketch of Sah's lemma.

Chapter V

Automorphisms

1. The Basic Diagram

This chapter, as its title indicates, concerns the group $\mathrm{Aut}(\pi)$ of automorphisms of a Bieberbach group π. A general reference is [20]. Much of this chapter is joint work with Han Sah and has never been published before.

As usual, we have the exact sequence

$$0 \longrightarrow M \longrightarrow \pi \xrightarrow{p} \Phi \longrightarrow 1. \tag{1}$$

Here we must be a bit more formal and let $j : \Phi \to \mathrm{Aut}(M)$ be the map induced by the action of Φ on M, so

$$[j(\sigma)]\,(m) = \bar{\sigma} m \bar{\sigma}^{-1}$$

for $m \in M$ and $\bar{\sigma} \in p^{-1}(\sigma)$. Let $\psi \in \mathrm{Aut}(\pi)$. By Proposition 4.1 of Chapter I (or Exercise 1.1 of Chapter III), M is the unique normal maximal abelian subgroup of π. Hence $\psi(M) = M$, so $\psi|M \in \mathrm{Aut}(M)$, and ψ induces a map ψ' on $\pi/M = \Phi$ as in the following diagram:

$$\begin{array}{ccccccccc} 0 & \longrightarrow & M & \longrightarrow & \pi & \xrightarrow{p} & \Phi & \longrightarrow & 1 \\ & & \downarrow{\scriptstyle \psi|M} & & \downarrow{\scriptstyle \psi} & & \downarrow{\scriptstyle \psi'} & & \\ 0 & \longrightarrow & M & \longrightarrow & \pi & \xrightarrow{p} & \Phi & \longrightarrow & 1. \end{array}$$

Define $F : \mathrm{Aut}(\pi) \to \mathrm{Aut}(M)$ and $G : \mathrm{Aut}(\pi) \to \mathrm{Aut}(\Phi)$ by $F(\psi) = \psi|M$ and $G(\psi) = \psi'$.

Proposition 1.1: *Let $\psi \in \mathrm{Aut}(\pi)$ and $\sigma \in \Phi$. Then*

$$j(\sigma) = F(\psi)^{-1} \circ j\left([G(\psi)]\,(\sigma)\right) \circ F(\psi). \tag{2}$$

Proof: Let $\bar{\sigma} \in p^{-1}(\sigma)$. Then $\forall m \in M$

$$\begin{aligned}[F(\psi) \circ j(\sigma)] \cdot m &= (\psi|M)(\bar{\sigma} m \bar{\sigma}^{-1}) \\ &= \psi'(\sigma) \cdot \psi(m) \\ &= [G(\psi)] \cdot [F(\psi)]\,(m) \\ &= [j\left([G(\psi)]\,(\sigma)\right) \circ F(\psi)]\,(m),\end{aligned}$$

as desired. □

Corollary 1.1: *Let N be the normalizer of $j(\Phi)$ in* $\mathrm{Aut}(M)$. *Then*

i) $imgF \subset N$ and

ii) $kerF \subset \ker G$.

Proof: Since M is a faithful Φ-module, j is injective, so ii) follows directly from equation (2), i.e. if $F(\psi) = 1$, then (2) becomes

$$j(\sigma) = j\left([G(\psi)]\,(\sigma)\right),$$

so $G(\psi) = 1$. Assertion i) follows even more directly from (2). □

Definition 1.1: Let $\mathrm{Aut}^0(\pi) = \ker F = \ker F \cap \ker G$ be those automorphisms of π which induce the identity on both M and Φ.

We now define an action, denoted by "$*$", of N, the normalizer of $j(\Phi)$ in $\mathrm{Aut}(M)$, on $H^2(\Phi; M)$ (actually the same definition works on $H^i(\Phi; M)$ for any i). Let $\varphi \in N$ and define $\varphi' \in \mathrm{Aut}(\Phi)$ by

$$\varphi'(\sigma) = j^{-1}\left[\varphi \cdot j(\sigma) \cdot \varphi^{-1}\right]$$

for $\sigma \in \Phi$. (j is injective since Φ is faithful.) Let $c : \Phi \times \Phi \to M$ be a 2-cocycle. Then we put

$$[\varphi * c](\sigma, \tau) = \varphi^{-1}\left[c\left(\varphi'(\sigma), \varphi'(\tau)\right)\right] \tag{3}$$

for $\sigma, \tau \in \Phi$.

Exercise 1.1: Show that equation (3) actually defines an action of N on $H^2(\Phi; M)$. (**Hint:** Show

$$[\delta(\varphi * c)](\sigma, \tau, \rho) = \varphi^{-1}\left[\delta c\left(\varphi'(\sigma), \varphi'(\tau), \varphi'(\rho)\right)\right]$$

for $\sigma, \tau, \rho \in \Phi$.)

Let $\alpha \in H^2(\Phi; M)$ be the cohomology class corresponding to the extension (1).

Lemma 1.1: *$F(\mathrm{Aut}(\pi)) = \mathrm{img}F$ is the isotropy subgroup of α in $H^2(\Phi; M)$, i.e.*

$$\mathrm{img}F = N_\alpha = \{\varphi \in N : \varphi * \alpha = \alpha\}.$$

Hence $\mathrm{Aut}(\pi)/\mathrm{Aut}^0(\pi)$ is isomorphic to N_α.

Exercise 1.2: Prove this lemma. (**Hint:** This is just a restatement of a special case of Theorem 2.2 of Chapter III.)

The next lemma can also be found in [39] or on page 119 of [45]. We use $\mathrm{Inn}(\pi)$ to denote the subgroup of $\mathrm{Aut}(\pi)$ of inner automorphisms, i.e. ψ is inner if there is $y \in \pi$ s.t. $\psi(x) = yxy^{-1}$ for all $x \in \pi$.

Lemma 1.2: *i) $\mathrm{Aut}^0(\pi)$ is isomorphic to $Z^1(\Phi; M)$, the group of 1-cocycles of Φ with values in M.*

ii) $\mathrm{Aut}^0(\pi) \cap \mathrm{Inn}(\pi)$ is isomorphic to $B^1(\Phi; M)$, the group of 1-coboundaries of Φ with values in M.

iii) $\mathrm{Aut}^0(\pi)/\mathrm{Aut}^0(\pi) \cap \mathrm{Inn}(\pi)$ is isomorphic to $H^1(\Phi; M)$.

Proof: Let $c \in Z^2(\Phi; M)$ be a representative of α. As usual, we think of π as the set $\Phi \times M$ with multiplication given by

$$(\sigma_1, m_1)(\sigma_2, m_2) = (\sigma_1\sigma_2, \sigma_1 \cdot m_2 + m_1 + c(\sigma_1, s_2)) \tag{4}$$

for $\sigma_1, \sigma_2 \in \Phi$ and $m_1, m_2 \in M$. Define a map $H : \mathrm{Aut}^0(\pi) \to C^1(\Phi; M)$ by the equation

$$\psi(\sigma, 0) = (\sigma, [H(\psi)](\sigma)) \tag{5}$$

for $\psi \in \mathrm{Aut}^0(\pi)$ and $\sigma \in \Phi$. In other words, we know that $\psi \in \mathrm{Aut}^0(\pi) \Rightarrow$ ψ is the identity on the first component, and $H(\psi)$ is that map from Φ to M that takes σ to the second component of $\psi(\sigma, 0)$.

Fix ψ and let $H(\psi) = h$. We have

$$\begin{aligned}\psi((\sigma,0)\cdot(\tau,0)) &= \psi(\sigma,0)\cdot\psi(\tau,0)\\ &= (\sigma, h(\sigma))\cdot(\tau, h(\tau))\\ &= (\sigma\tau, \sigma\cdot h(\tau) + h(\sigma) + c(\sigma,\tau)).\end{aligned}$$

On the other hand,

$$\begin{aligned}\psi((\sigma,0)\cdot(\tau,0)) &= \psi(\sigma\tau, c(\sigma,\tau))\\ &= \psi((1, c(\sigma,\tau)\cdot(\sigma\tau,0)) \qquad \text{since } c(1,\sigma\tau) = 0\\ &= \psi(1, c(\sigma,\tau))\cdot\psi(\sigma\tau,0)\\ &= (\,c(\sigma,\tau))\cdot(\sigma\tau, h(\sigma\tau)) \qquad \text{since } \psi \in \mathrm{Aut}^0(\pi)\\ &= (\sigma\tau, h(\sigma\tau) + c(\sigma,\tau)).\end{aligned}$$

Hence

$$\sigma \cdot h(\tau) + h(\sigma) + c(\sigma,\tau) = h(\sigma\tau) + c(\sigma,\tau),$$

so $\delta h = 0$ and $h \in Z^1(\Phi; M)$, and we see that H is a map from $\mathrm{Aut}^0(\pi)$ to $Z^1(\Phi; M)$.

Exercise 1.3: Show that H is a homomorphism.

Suppose that $H(\psi) = 0$, i.e. $[H(\psi)](\sigma) = 0\ \forall\sigma \in \Phi$. Then (5) shows that ψ is the identity, so H is injective. On the other hand, let $h \in Z^1(\Phi; M)$ be given and define $\psi \in \mathrm{Aut}^0(\pi)$ by

$$\psi(\sigma, m) = (\sigma, m + h(\sigma)).$$

Exercise 1.4: Show that $\psi \in \mathrm{Aut}^0(\pi)$.

Therefore H is surjective and i) follows.

For ii), suppose $\psi \in \mathrm{Aut}^0(\pi) \cap \mathrm{Inn}(\pi)$, i.e. suppose ψ is conjugation by (σ_0, a_0). Since $\psi|M$ is the identity, we must have

$$\begin{aligned}(\sigma_0, a_0)(1,m)(\sigma_0,a_0)^{-1} &= (\sigma_0, \sigma_0 \cdot m + a_0)(\sigma_0^{-1}, -\sigma_0^{-1}a_0 - c(\sigma_0,\sigma_0^{-1}))\\ &= (1, -a_0 - \sigma_0 c(\sigma_0,\sigma_0^{-1}) + \sigma_0 \cdot m + a_0\\ &\qquad + c(\sigma_0,\sigma_0^{-1}))\\ &= (1, (1-\sigma_0)c(\sigma_0,\sigma_0^{-1}) + \sigma_0 \cdot m)\\ &= (1, m).\end{aligned}$$

This says that $(1-\sigma_0)c(\sigma_0,\sigma_0^{-1}) = (1-\sigma_0)\cdot m$ and since m is arbitrary in M, we must have $\sigma_0 = 0$, so ψ is conjugation by $(1, a_0)$. From (5) we get

$$\begin{aligned}\psi(\sigma, 0) &= (\sigma, [H(\psi)](\sigma))\\ &= (1,a_0)(\sigma,0)(1,a_0)^{-1}\\ &= (\sigma, a_0)(1, -a_0)\\ &= (\sigma, -\sigma\cdot a_0 + a_0),\end{aligned}$$

so

$$[H(\psi)](\sigma) = a_0 - \sigma \cdot a_0$$

for some $a_0 \in M$ which precisely says that $H(\psi) \in B^1(\Phi; M)$.

Exercise 1.5: Prove the rest of ii), i.e. show that H restricted to $\mathrm{Aut}^0(\pi) \cap \mathrm{Inn}(\pi)$ is a bijection onto $B^1(\Phi; M)$.

iii) follows directly from i) and ii). □

Lemma 1.3: *Define $\varphi : M \to B^1(\Phi; M)$ by $[\varphi(m)](\sigma) = m - \sigma \cdot m$. Then the sequence*

$$0 \longrightarrow M^\Phi \longrightarrow M \xrightarrow{\varphi} B^1(\Phi; M) \longrightarrow 0$$

is exact, where $M^\Phi = \{m \in M : \sigma \cdot m = m \ \forall \sigma \in \Phi\}$.

Exercise 1.6: Prove this lemma.

Recall that the group of *outer automorphisms* is defined by

$$\mathrm{Out}(\pi) = \mathrm{Aut}(\pi)/\mathrm{Inn}(\pi).$$

Thus an outer automorphism is not an automorphism at all, but the nomenclature is too well-known to change now. Also recall that N is the normalizer of $j(\Phi)$ in $\mathrm{Aut}(M)$, and N_α is the subgroup of N of those elements that leave $\alpha \in H^2(\Phi; M)$ fixed. Before we state the main theorem of this section, we need a proposition which is of interest itself. First note that we can regard Φ as a subgroup of N via j.

Proposition 1.2: *The action of Φ on $H^2(\Phi; M)$ given by (3) is trivial. Hence $\Phi \subset N_\alpha$.*

Proof: For $\lambda \in \Phi$ and $c \in Z^2(\Phi; M)$, define a 1-cochain $g : \Phi \to M$ by

$$g(\rho) = c(\lambda^{-1}, \lambda\rho\lambda^{-1}) - c(\rho, \lambda^{-1})$$

for $\pi \in \Phi$. Then

$$\begin{aligned}(\lambda * c)(\sigma, \tau) &= \lambda^{-1} \cdot c(\lambda\sigma\lambda^{-1}, \lambda\tau\lambda^{-1})\\ &= c(\sigma\lambda^{-1}, \lambda\tau\lambda^{-1}) - c(\lambda^{-1}, \lambda\sigma\tau\lambda^{-1}) + c(\lambda^{-1}, \lambda\sigma\lambda^{-1})\\ &\qquad\qquad \text{since } \delta c(\lambda^{-1}, \lambda\sigma\lambda^{-1}, \lambda\tau\lambda^{-1}) = 0\\ &= \delta g(\sigma, \tau) + \sigma \cdot c(\tau, \lambda^{-1}) - c(\sigma\tau, \lambda^{-1}) + c(\sigma, \tau\lambda^{-1})\\ &= \delta g(\sigma, \tau) + c(\sigma, \tau) \qquad \text{since } \delta c(\sigma, \tau, \lambda^{-1}) = 0\end{aligned}$$

which is exactly what we wanted. □

Exercise 1.7: Make the obvious extension of (3) to get an action of N on $H^i(\Phi; M)$ and show that the subgroup Φ of N acts trivially here too.

We should, of course, be saying "the subgroup $j(\Phi)$ of N," but we will continue with the "abuse of terminology." Notice that Φ is normal in N and N_α.

Theorem 1.1 : *The following is a commutative diagram in which all rows and columns are exact:*

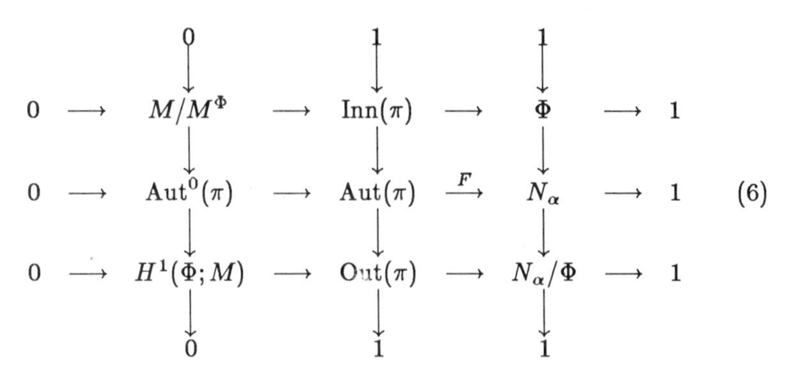

$$\begin{array}{ccccccccc} & & 0 & & 1 & & 1 & & \\ & & \downarrow & & \downarrow & & \downarrow & & \\ 0 & \longrightarrow & M/M^{\Phi} & \longrightarrow & \mathrm{Inn}(\pi) & \longrightarrow & \Phi & \longrightarrow & 1 \\ & & \downarrow & & \downarrow & & \downarrow & & \\ 0 & \longrightarrow & \mathrm{Aut}^0(\pi) & \longrightarrow & \mathrm{Aut}(\pi) & \xrightarrow{F} & N_\alpha & \longrightarrow & 1 \\ & & \downarrow & & \downarrow & & \downarrow & & \\ 0 & \longrightarrow & H^1(\Phi; M) & \longrightarrow & \mathrm{Out}(\pi) & \longrightarrow & N_\alpha/\Phi & \longrightarrow & 1 \\ & & \downarrow & & \downarrow & & \downarrow & & \\ & & 0 & & 1 & & 1 & & \end{array} \qquad (6)$$

In particular, $\mathrm{Out}(\pi)$ *contains a subgroup isomorphic to* $H^1(\Phi; M)$ *whose quotient is isomorphic to* N_α/Φ.

Proof: The commutativity is easy, and we leave it as an exercise. For the exactness, we start with the middle row. Lemma 1.1 says that F is surjective, and the definition of $\mathrm{Aut}^0(\pi)$ says it is the kernel of F. Thus the middle row is exact.

The middle column is exact by definition and similarly for the right column. The left column is exact by Lemmas 1.2 and 1.3.

Now we look at the top row. $F(\psi) = \psi|M$ for $\psi \in \mathrm{Aut}(\pi)$. If ψ is conjugation by (σ, m), then since $\delta c(1, \sigma, \sigma^{-1}) = 0$, $c(\sigma, \sigma^{-1}) = 0$, so

$$\begin{aligned}(\sigma, m)(1, m')(\sigma^{-1}, -\sigma^{-1} \cdot m) &= (1, -m + m + \sigma \cdot m' + c(\sigma, \sigma^{-1})) \\ &= (1, \sigma \cdot m),\end{aligned}$$

and we see that $[F(\psi)](m') = \sigma \cdot m'$. Hence F maps $\mathrm{Inn}(\pi)$ onto Φ. It follows from the above that if conjugation by (σ, m) is in the kernel of F, then $\sigma = 1$. On the other hand,

$$(1, m)(\sigma', m')(1, -m) = (\sigma', (1 - \sigma')m + m')$$

shows that conjugation by $(1, m)$ induces the identity automorphism iff $m \in M^{\Phi}$, so the kernel of $F|\mathrm{Inn}(\pi)$ is isomorphic to M/M^{Φ}. Hence the top row is exact.

It only remains to define the maps in the bottom row and show it is exact, and this follows from the following general exercise which is an easy "diagram chase" and, in fact, is a part of Exercise 3.4 of Chapter III.

Exercise 1.8: Let

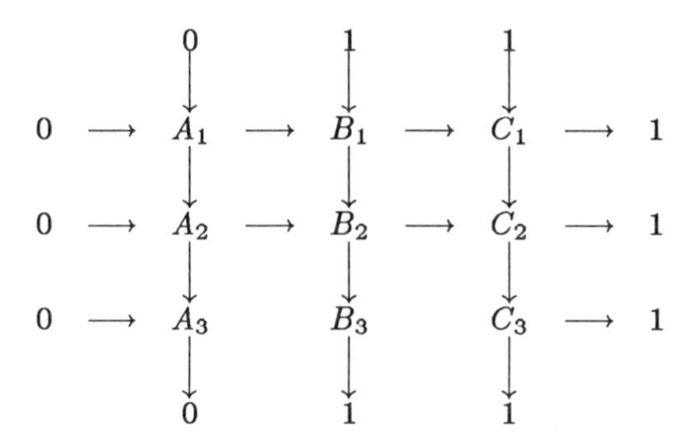

be a commutative diagram of groups (with A_i abelian for $i = 1, 2, 3$) s.t. the rows and columns are exact. Show that there are unique maps from A_3 to B_3 and B_3 to C_3 which make the resulting diagram commutative. Show that for these maps, the bottom row is exact.

This finishes the proof of the theorem. □

Remark: We call diagram (6) the *basic diagram*. The top row of the basic diagram gives us a pretty good hold on $\mathrm{Inn}(\pi)$, so to understand $\mathrm{Aut}(\pi)$, what we need to look at is $\mathrm{Out}(\pi)$. $\mathrm{Out}(\pi)$ also has an interesting geometric intrepretation in the context of flat manifolds (see Section 6). It is clear how N_α/Φ acts on $H^1(\Phi; M))$, but what is the cohomology class in $H^2(N_\alpha/\Phi; H^1(\Phi; M)))$ that corresponds to the extension given by the bottom row? The next four sections are devoted to this question which will involve us in some quite interesting and important mathematics. In particular, you should begin see where the complicated idea of a spectral sequence comes from.

2. The Hochschild–Serre Exact Sequence

Suppose we have a group extension

$$1 \longrightarrow K \longrightarrow G \longrightarrow Q \longrightarrow 1 \tag{7}$$

where this time we will *not* assume that K is abelian. Of course, we must assume that K is normal in G. Suppose we have an abelian group A which, to make things a little simpler at first, we consider as a trivial K-module, a trivial G-module, and a trivial Q-module. We can then ask how $H^*(G; A)$ is related to $H^*(K; A)$ and $H^*(Q; A)$. We can ask, but in general, the answer is very complex. There is a very, very complicated object called a *spectral sequence* which reduces the problem of computing $H^*(G;A)$ to a sequence of cohomology computations. We hope to discuss this situation in the next volume.

It turns out, however, if you are only interested in $H^1(G;A)$ and $H^2(G;A)$, there is an exact sequence with six terms that gives some information. The neat way to derive this exact sequence is as a trivial consequence of the spectral sequence mentioned above. In this section we will give an elementary derivation which will perhaps give some indication of the difficulties of the problem which lead to the construction of the spectral sequence.

First we need to describe some homomorphisms between cohomology groups. We assume we have a group G with a normal subgroup K which need not be abelian. We call the quotient group Q. This yields the exact sequence (7). Let A be a G-module. We do not assume that A is a trivial G-module here. Clearly A is also a K-module.

Definition 2.1: We define the *restriction* (from G to K) to be the homomorphism

$$\operatorname{res}_K^G : H^i(G;A) \to H^i(K;A)$$

induced by restricting a cocycle $c \in C^i(G;A)$ to K. If there is no danger of confusion, we denote res_K^G by res. This definition actually works even if K is not normal in G.

Exercise 2.1: i) Show that res_K^G is nothing more than the map i^* where $i : K \hookrightarrow G$ is the inclusion map.

ii) Show that res is "natural" in A, i.e. if $f : A \to B$ is a homomorphism, the diagram

$$\begin{array}{ccc} H^n(G;A) & \xrightarrow{\text{res}} & H^n(K;A) \\ \downarrow f_* & & \downarrow f_* \\ H^n(G;B) & \xrightarrow{\text{res}} & H^n(K;B) \end{array}$$

is commutative.

iii) Suppose L is normal in K. Show that $\operatorname{res}_L^K \circ \operatorname{res}_K^G = \operatorname{res}_L^G$.

Now recall that $A^K = \{a \in A : k \cdot a = a \quad \forall k \in K\}$. Let $p : G \to Q$ be the projection. Notice that A^K is a Q-module because K acts trivially on it.

Definition 2.2: We define the *inflation* (from Q to G) to be the homomorphism

$$\inf_G^Q : H^i(Q; A^K) \to H^i(G; A)$$

induced by the formula

$$[\inf_G^Q(c)](g_1, \ldots, g_i) = c(p(g_1), \ldots, p(g_i))$$

where $c \in C^i(Q; A^K)$ and $g_k \in G$ for $k = 1, \ldots, i$. If there is no danger of confusion, we denote $\inf_G^Q$ merely by inf. This construction, of course, requires K to be normal in G.

Exercise 2.2: i) Show that inf is well-defined on cohomology.
ii) Show that inf is "natural" in A.

The next homomorphism we need has a more complicated definition. For this we must require that K be normal in G and that A be a G-module on which K acts trivially, so Q acts naturally on A. Since K is normal in G, G acts by conjugation on K, and we can let G act on $H^i(K; A)$ by the following formula:

$$(g * c)(k_1, \ldots, k_i) = g \cdot c(g^{-1}k_1g, \ldots, g^{-1}k_ig) \tag{8}$$

for $c \in C^i(K; A), g \in G$, and $k_j \in K$.

We now want to define a connecting map

$$\mathrm{con}_Q^K : H^1(K; A)^G \to H^2(Q; A).$$

To do this let $a \in C^2(Q; K)$ be a 2-cocycle in the cohomology class corresponding to the extension (7). Let $c \in C^1(K; A)$ be a 1-cocycle representing a class $[c] \in H^1(K; A)^G$. Then we let $\mathrm{con}([c])$ be the class of the 2-cocycle $z \in C^2(Q; A)$ defined by

$$z(q_1, q_2) = c(a(q_1, q_2)). \tag{9}$$

To see that this makes sense we must show that z is indeed a 2-cocycle and that its cohomology class is independent of the choice of a and c. We are aided in our task by the fact that c is a very nice cochain indeed.

We will show

$$c(k_1k_2) = c(k_1) + c(k_2), \tag{10}$$

and

$$c(gkg^{-1}) = g \cdot c(k). \tag{11}$$

It is easy to see that (10) is true since

$$\delta c(k_1, k_2) = k_1 c(k_2) - c(k_1 k_2) + c(k_1) = c(k_2) - c(k_1 k_2) + c(k_1) = 0$$

since A is a trivial K-module.

To see that (11) holds, notice that we have assumed that $[c]$ is in $H^1(K; A)^G$, so $g * c$ and c are cohomologous, i.e. there is $b \in C^0(K; A)$ s.t.

$$(g * c)(k) - c(k) = (\delta b)(k) = k \cdot b - b \tag{12}$$

for any $k \in K$. But b is just an element of A which is a trivial K-module, so $k \cdot b = b$ and $(g * c)(k) = c(k)$, i.e.

$$g \cdot c(g^{-1}kg) = c(k),$$

which yields (12).

Now it is easy to see that $\delta z = 0$. In fact, using (10), (11), and (12), we get

$$\begin{aligned}\delta z(q_1, q_2, q_3) &= q_1 \cdot z(q_2, q_3) - z(q_1 q_2, q_3) + z(q_1, q_2 q_3) - z(q_1, q_2) \\ &= c\,(q_1 \cdot a(q_2, q_3) - a(q_1 q_2, q_3) + a(q_1, q_2 q_3) - a(q_1 q_2)) \\ &= c\,(\delta a(q_1, q_2, q_3)) = 0.\end{aligned}$$

Using (10), we can see that to show that the cohomology class of z is independent of the choice of a it suffices to show that for any $b \in C^1(Q; K)$ we can find $b' \in C^1(Q; K)$ s.t. $c \circ \delta b = \delta b'$. Now

$$c\,(q_1 \cdot b(q_2) = b(q_1 q_2) + b(q_1)) = q_1 \cdot c(b(q_2)) - c(b(q_1 q_2)) + c(b(q_1))$$

so we can take $c \circ b$ for b'. It is trivial from (10) to see that the class of z is independent of c. Finally, using the terminology of [56], we can make the following

Definition 2.3: We define the *connection* (from K to Q) to be the homomorphism

$$\mathrm{con}_Q^K : H^1(K; A)^G \to H^2(Q; A)$$

induced by the formula

$$\left[\operatorname{con}_Q^K(c)\right](q_1, q_2) = c(a(q_1, q_2))$$

for $c \in C^1(K; A)$ with $[c] \in H^1(K; A)^G$ and $q_i \in Q$. If there is no danger of confusion, we denote con_Q^K merely by con.

Remark: The connection is sometimes called the *transgression.* It is a special case of a more general notion that arises in the study of spectral sequences. One must not confuse this with the notion of connection introduced in Chapter II, but since they are so different, it is unlikely this will happen.

Before we state the main theorem of this section, we need a lemma. Recall

$$\operatorname{res} : H^i(G; A) \to H^i(K; A).$$

Lemma 2.1: *If K is normal in G, then the image of res lies in $H^i(K; A)^G$.*

Proof: We do the case $i = 1$ which is all we will use. Let $c \in Z^1(G; A)$, so

$$g_1 \cdot c(g_2) - c(g_1 g_2) + c(g_1) = 0 \tag{13}$$

for any $g_1, g_2 \in G$. Since there are no 1-coboundaries (remember A is a trivial K-module), we must show that

$$g * c = c \qquad \forall g \in G.$$

First take $g_1 = g$ and $g_2 = g^{-1}kg$ for some $k \in K$. Then (13) yields

$$g \cdot c(g^{-1}kg) - c(kg) + c(g) = 0. \tag{14}$$

Now take $g_1 = k$ and $g_2 = g$. Then (13) yields

$$c(g) = c(kg) - c(k) \tag{15}$$

since K acts trivially on A. Combining (14) and (15), we get $g * c = c$ as desired. □

Exercise 2.3: Prove the lemma for $i > 1$.

Theorem 2.1 (Hochschild–Serre): *Let G be a group with a normal subgroup K, let A be a G-module on which K acts trivially, and let $Q = G/K$. Then the following sequence is exact:*

$$0 \longrightarrow H^1(Q;A) \xrightarrow{\text{inf}} H^1(G;A) \xrightarrow{\text{res}} H^1(K;A)^G \xrightarrow{\text{con}} H^2(Q;A) \xrightarrow{\text{inf}} H^2(G;A).$$

Proof: Since there are no 1-coboundaries, it is trivial to see that

$$\text{inf} : H^1(Q;A) \to H^1(G;A)$$

is injective. It is also trivial to see that res $\circ$ inf $= 0$. Similarly if $c \in Z^1(G;A)$ and $c(k) = 0\ \forall k \in K$, then since $\delta c = 0$, we get $c(kg) = c(g)$, so c defines a map $c' : Q \to A$. It is easy to see that c' is a 1-cocycle (i.e. in $Z^1(Q;A)$) and $\text{inf}(c') = c$. Thus we have seen that the sequence is exact at $H^1(Q;A)$ and $H^1(G;A)$.

Now suppose that $c \in Z^1(G;A)$ and consider con $\circ$ res$([c])$. We want to show that it is cohomologous to 0, i.e. we want $b \in C^1(Q;A)$ s.t.

$$c(a(q_1,q_2)) = (\delta b)(q_1,q_2).$$

We know that c is defined on all of G and that $\delta c = 0$. Set

$$b(q) = c(s(q))$$

where $s : Q \to G$ is a section (not a homomorphism) with

$$a(q_1,q_2) = s(q_1)s(q_2)s(q_1q_2)^{-1}$$

(see the discussion after Exercise 5.8 of Chapter I). Now

$$\begin{aligned}
\delta b(q_1,q_2) &= q_1 \cdot c\,(s(q_1)) - c((s(q_1q_2)) + c\,(s(q_1)) \\
&= [c\,(s(q_1)s(q_2)) - c\,(s(q_1))] - c((s(q_1q_2)) + c\,(s(q_1)) \\
&= c((s(q_1)s(q_2)s(q_1q_2)^{-1})\,.
\end{aligned}$$

To get the last equality, note that

$$\delta c((s(q_1)s(q_2)s(q_1q_2)^{-1}, s(q_1q_2)) = 0,$$

and since $s(q_1)s(q_2)s(q_1q_2)^{-1} \in K$, it acts trivially on A. Thus $\delta b(q_1,q_2) = c\,(a(q_1,q_2))$ as desired, and we see that con$\circ$res$= 0$.

Now assume that $[c] \in H^1(K;A)^G$ and that $\mathrm{con}([c]) = 0$. We show that $[c]$ is in the image of res. We have $c \in Z^1(K;A)$ and $g * [c] = [c]$, but since there are no 1-coboundaries, we have $g * c = c$. We also know that $\mathrm{con}([c]) = 0$ so there is $b \in C^1(Q;A)$ s.t.

$$c\left(a(q_1, q_2)\right) = \delta b(q_1, q_2). \tag{16}$$

We want to find $\bar{c} \in Z^1(G;A)$ s.t. $\bar{c}|K = c$. We are going to think of $g \in G$ as consisting of two pieces, $p(g) \in Q$ (recall $p : G \to Q$ is the projection) and $g \cdot s\left(p(g)\right)^{-1} \in K$. We use c on K and b on Q and define

$$\bar{c}(g) = c\left(s \cdot s\left(p(g)\right)^{-1}\right) + b\left(p(g)\right) \in A. \tag{17}$$

Clearly $\bar{c}(k) = c(k)\ \forall k \in K$. What needs to be shown is that $\delta\bar{c} = 0$. Before we begin, let's record some facts about c. $\delta c = 0$ is equivalent to

$$c(k_1 k_2) = c(k_1) + c(k_2) \tag{18}$$

since K acts trivially on A. Now K is not necessarily abelian, but (18) implies

$$c(k_1 k_2) = c(k_2 k_1). \tag{19}$$

Also $g * c = c$ can be written

$$c(gkg^{-1}) = g \cdot c(k). \tag{20}$$

Exercise 2.4: Show that $\delta\bar{c} = 0$. (**Hint:** This is a non-trivial computation. If it were not so hard to type in TEX, we would have included it in the text. What you want to do is first to use (17) and write out the ensuing six terms. Then use (18), (19), and (20) to get what you want. The key equation is (19). You should group things so as to get elements as arguments of c which are in K, and can therefore be permuted. It took us about twelve lines to get it. Good luck!)

Hence (?) the sequence is exact at $H^1(K;A)$.

We now want to show that $\inf \circ \mathrm{con} = 0$. Let $c \in Z^1(K;A)$. We want to find $b \in C^1(G;A)$ s.t.

$$\delta b(g_1, g_2) = c\left(a\left(p(g_1), p(g_2)\right)\right). \tag{21}$$

Since $[c] \in H^1(K;A)^G$, (18), (19), and (20) hold for this c too. Using ideas similar to those in the last computation, we put

$$b(g) = -c\left(g \cdot s\left(p(g)\right)^{-1}\right).$$

Then we can get (21) exactly as above.

Finally let's take $c \in Z^1(Q;A)$ and suppose $\inf([c]) = 0$. We show that $[c]$ is in the image of con. Since $\inf([c]) = 0$, there is $b \in C^1(G;A)$ s.t.

$$c\left(p(g_1), p(g_2)\right) = \delta b(g_1, g_2). \tag{22}$$

We need to find a map $h \in C^1(K;A)$ s.t. $[h] \in H^1(K;A)^G$ and $\operatorname{con}([h]) = [c]$. Let's try $h = -b|K$. Then $\operatorname{con}([h])$ is represented by the map $b \circ a : Q \times Q \to A$. We must show that $\delta(b|K) = 0$ and $g * (b|K) = (b|K)$. The first fact about the coboundary follows immediately from (22). In fact, (22) says that $\delta b(g_1, g_2) = 0$ if either g_1 or g_2 is in K. Hence

$$\delta b(g, gkg^{-1}) = p(g) \cdot b(g^{-1}kg) - b(kg) + b(g) = 0,$$

and

$$\delta b(k, g) = b(g) - b(kg) + b(k) = 0$$

for $g \in G$ and $k \in K$. These last two equations yield the equation

$$p(g) \cdot b(g^{-1}kg) = b(g)$$

which says $(g * b)(k) = b(k)$ for $g \in G$ and $k \in K$.

The very last thing to show is that $\operatorname{con}([h]) = [c]$. For this we must again compute. What we must compute is

$$b\left(a(q_1, q_2)\right) = b\left(s(q_1)s(q_2)s(q_1q_2)^{-1}\right).$$

Notice that (22) implies that

$$b(g_1g_2) = g_1 \cdot b(g_2) + b(g_1) - c\left(p(g_1), p(g_2)\right). \tag{23}$$

Exercise 2.5: Show that

$$b\left(a(q_1, q_2)\right) = q_1 \cdot b\left(s(q_2)\right) - b\left(s(q_1q_2)\right) + b\left(s(q_1)\right) - c(q_1, q_2). \tag{24}$$

(**Hint:** This computation is not as bad as the last omitted one. Use (23) a number of times, and you will get it.)

On the one hand, equation (24) doesn't look so hot. On the other hand, we never should have expected $-b\left(a(q_1,q_2)\right)$ to equal $c(q_1,q_2)$ since we can alter b by a coboundary and (22) will still hold. The point is that we only need $-b\circ a$ to be in the *class* of c. To see that this is true, let $z\in C^1(G;A)$ be given by

$$z(g)=b\left(s\left(p(g)\right)\right).$$

Then

$$\delta z(g_1,g_2)=g_1\cdot b\left(s\left(p(g_2)\right)\right)-b\left(s\left(p(g_1g_2)\right)\right)+b\left(s\left(p(g_1)\right)\right),$$

or

$$\delta z\left(s(q_1),s(q_2)\right)=q_1\cdot b\left(s(q_2)\right)-b\left(s(q_1q_2)\right)+b\left(s(q_1)\right).$$

So we have

$$b\left(a(q_1,q_2)\right)=\delta z((s(q_1),s(q_2))-c(q_1,q_2),$$

and $\mathrm{con}([b])=-[c]$ as desired. $\square$

Remarks: i) It should now be clear that it will be difficult to continue the exact sequence any further. Even if there were a way to do it, it would be exceedingly complex. It turns out that there is no way to do it. You can map $H^2(G;K)$ to $H^2(K;A)^G$ by the restriction map, res, but the sequence will not be exact there. Roughly speaking, $H^1(G;A)$ is made up from $H^1(Q;A)$ and $H^1(K;A)^G$, while $H^2(G;A)$ not only involves $H^2(Q;A)$ and $H^2(K;A)^G$, but also a third group which turns out to be $H^1(Q;H^1(K;A))$. In addition, these cohomology groups are not directly involved themselves, but what is used are quotient groups of subgroups of them ("subquotients"). The situation is even more complicated for $H^3(G;A)$. The device which handles all this complication is the spectral sequence, the Hochschild–Serre spectral sequence to be exact.

ii) Since we have not assumed that K is abelian, we don't have an action of Q on K, and so we have used the action of G on K given by conjugation. However if K happens to be abelian, we do get an action of Q on K and actually of Q on $H^i(K;A)$. In this case the theorem works with $H^1(K;A)^G$ replaced by $H^1(K;A)^Q$.

iii) Suppose we did not assume that K acts trivially on A. Then we still get an exact sequence, but we must replace $H^i(Q;A)$ with $H^i(Q;A^K)$

$(i = 1, 2)$, so now the sequence is

$$0 \longrightarrow H^1(Q; A^K) \xrightarrow{\text{inf}} H^1(G; A) \xrightarrow{\text{res}} H^1(K; A)^G \xrightarrow{\text{con}} H^2(Q; A^K) \xrightarrow{\text{inf}} H^2(G; A). \tag{25}$$

Exercise 2.6: Prove parts i) and ii) of the above remark.

3. 9-Diagrams

Diagram (6) is an example of what is defined below to be a "9-diagram." Such diagrams have been occurring in various branches of mathematics recently (see [26] for example). In this section we prove some basic material about 9-diagrams which will be useful in studying our basic 9-diagram (6).

Definition 3.1: A *9-diagram* is a commutative diagram of groups and homomorphisms of the following form:

$$\begin{array}{ccccccccc} & & 0 & & 1 & & 1 & & \\ & & \downarrow & & \downarrow & & \downarrow & & \\ 0 & \longrightarrow & A_1 & \xrightarrow{i_1} & B_1 & \xrightarrow{p_1} & C_1 & \longrightarrow & 1 \\ & & \downarrow{\scriptstyle i_A} & & \downarrow{\scriptstyle i_B} & & \downarrow{\scriptstyle i_C} & & \\ 0 & \longrightarrow & A_2 & \xrightarrow{i_2} & B_2 & \xrightarrow{p_2} & C_2 & \longrightarrow & 1 \\ & & \downarrow{\scriptstyle p_A} & & \downarrow{\scriptstyle p_B} & & \downarrow{\scriptstyle p_C} & & \\ 0 & \longrightarrow & A_3 & \xrightarrow{i_3} & B_3 & \xrightarrow{p_3} & C_3 & \longrightarrow & 1 \\ & & \downarrow & & \downarrow & & \downarrow & & \\ & & 0 & & 1 & & 1 & & \end{array} \tag{26}$$

where the groups A_i for $i = 1, 2, 3$ are abelian, and all the rows and columns are exact. There is one additional requirement. Recall that A_2 and consequently A_1 are, in the usual manner, C_2-modules.

Exercise 3.1: Show that A_3 is a C_2-module in the obvious way.

Our requirement is that the subgroup C_1 of C_2 act trivially on A_3.

Remark: One can think of a 9-diagram as an "extension of group extensions." In fact, if you want to be fancy, you could define a category of "supergroups," where the objects are short exact sequences of groups, and then the 9-diagram (26) would tell us that the supergroup B is an extension of the abelian supergroup A by the supergroup C.

The first problem we will investigate is the following: Suppose we have a short exact sequence of groups

$$1 \longrightarrow C_1 \xrightarrow{i_C} C_2 \xrightarrow{p_C} C_3 \longrightarrow 1 \tag{27}$$

and a short exact sequence of C_2-modules

$$0 \longrightarrow A_1 \xrightarrow{i_A} A_2 \xrightarrow{p_A} A_3 \longrightarrow 0 \tag{28}$$

s.t. C_1 acts trivially on A_3 so that A_3 is naturally a C_3-module. We want to find all short exact sequences of groups

$$1 \longrightarrow B_1 \xrightarrow{i_B} B_2 \xrightarrow{p_B} B_3 \longrightarrow 1 \tag{29}$$

which together with appropriate homomorphisms $i_j : A_j \to B_j$ and $p_j : B_j \to C_j$ $(i = 1, 2, 3)$ make up a 9-diagram like (26).

Suppose we have a 9-diagram (26). Then the rows are exactly the kind of group extension we know and love. As usual B_i determines a cohomology class $\alpha_i \in H^2(C_i; A_i)$ and if $c_i \in \alpha_i$, we can consider the groups B_i as the sets $C_i \times A_i$ with multiplication given by

$$(\gamma, a)(\delta, b) = (\gamma\delta, \gamma \cdot a + b + c_i(\gamma, \delta))$$

for $\gamma, \delta \in C_i$ and $a, b \in A_i$. As usual we sometimes identify A_i with $\{(1, a) : a \in A_i\} \subset B_i$, and then we will write A_i multiplicatively instead of the usual additive notation for the abelian group A_i.

Lemma 3.1: *We can find a section $s_2 : C_2 \to B_2$ s.t.*

i) $s_2(\gamma) \in B_1$ $\forall \gamma \in C_1$,

ii) if $\sigma : C_3 \to C_2$ is any section, then

$$s_2(\gamma \cdot \sigma(\delta)) = s_2(\gamma) \cdot s_2(\sigma(\delta))$$

for $\gamma \in C_1$ and $\delta \in C_3$, and

iii) $s_2(\gamma\beta)B_1 = s_2(\beta)B_1$ for $\gamma \in C_1$ and $\beta \in C_2$.

Furthermore each such section yields a section $s_3 : C_3 \to B_3$ by setting

$$s_3(\gamma_3) = p_B(s_2(\gamma_2)) \qquad \text{where } p_C(\gamma_2) = \gamma_3, \tag{30}$$

and all sections from c_3 to B_3 can be obtained in this manner.

Exercise 3.2: Prove this lemma. (This should enable you to feel at home in a 9-diagram. **Hint:** Let $s_1 : C_1 \to B_1$, $s_3 : C_3 \to B_3$, $\sigma_C : C_3 \to C_2$, and $\sigma_B : B_3 \to B_2$ be sections. Then for $\gamma \in C_2$ put $s_2(\gamma) = s_1(\gamma\sigma_C(p_C(\gamma))^{-1}) \cdot \sigma_B(s_3(p_C(\gamma)))$.)

We now assume we have sections $s_i : C_i \to B_i$ s.t. $s_1 = s_2|C_1$, s_2 and s_3 satisfy (30), and s_2 satisfies the conditions of Lemma 3.1. Furthermore we assume our 2-cocycles c_i are given by

$$c_i(\gamma, \delta) = s_i(\gamma)s_i(\delta)s_i(\gamma\delta)^{-1}. \tag{31}$$

for $\gamma, \delta \in C_i$.

Lemma 3.2: *The cocycles c_i satisfy*

i) $c_2(\gamma, \delta)$ and $c_2(\delta, \gamma)$ are elements of A_1 if $\gamma \in C_1$ for any $\delta \in C_2$,

ii) $c_1 = \mathrm{res}(c_2) = c_2|C_1$, and

iii) given $\gamma_3, \delta_3 \in C_3$ and $\gamma_2, \delta_2 \in C_2$ with $p_C(\gamma_2) = \gamma_3$ and $p_C(\delta_2) = \delta_3$, then

$$c_3(\gamma_3, \delta_3) = c_2(\gamma_2, \delta_2)A_1.$$

Exercise 3.3: Prove this lemma.

Proposition 3.1: *Suppose we have sequences of groups (27) and (28) (satisfying, of course, that A_3 is a trivial C_1-module). Then each sequence (29) which fits into the 9-diagram (6) corresponds to classes $\alpha_i \in H^2(C_i; A_i)$ s.t.*

i) $\mathrm{res}(\alpha_2) = (i_A)_*(\alpha_1)$,

ii) $(p_A)_*(\alpha_2) = \inf(\alpha_3)$, and

iii) $\alpha_1 \in H^2(C_1; A_1)^{C_2}$.

Proof: i) follows from ii) of Lemma 3.1, while ii) follows from iii) of Lemma 3.1. iii) follows from i) and Lemma 2.1 and uses the fact that B_1 is normal in B_2. □

Remark: If we are given the A_i's and the C_i's and want to find some B_i's, Proposition 3.1 tells us to look for α_1, α_2, and α_3 satisfying i), ii), and iii). We might suppose that if we have α_1 and α_2 satisfying i) and iii), we can find an α_2 satisfying ii), but this is not the case, because i) and iii) will not ensure that B_1 will be normal in B_2 (see the example below). However, if we have α_2 and α_3 satisfying ii), we *can* find α_1 satisfying i) and iii), and hence get a sequence of B_i filling in the 9-diagram (see Theorem 3.1 below).

Example 3.1: This is a example of an extension that satisfies i) and iii), but with no α_2 satisfying ii). Recall that D_8, the *dihedral group of order eight*, is generated by two elements a and b with the relations $a^4 = 1$, $b^2 = 1$, and $bab = a^3 = a^{-1}$.

Exercise 3.4: i) Show that D_8 has eight elements, namely, $1, a, a^2, a^3, b, ba, ba^2$, and ba^4. Write out the group table for D_8.

ii) Show that D_8 is generated by two elements of order two.

iii) Show that D_8 satisfies an exact sequence

$$0 \longrightarrow \mathbb{Z}_4 \longrightarrow D_8 \longrightarrow \mathbb{Z}_2 \longrightarrow 0.$$

Compute the action of the quotient $\mathbb{Z}_2$ on the kernel $\mathbb{Z}_4$. Also, compute $H^2(\mathbb{Z}_2; \mathbb{Z}_4)$ with this action and determine which cohomology class corresponds to this extension.

iv) Show that the subgroup generated by a^2 is in the center Z of D_8. Let $p_2 : D_8 \to D_8/Z$ be the canonical projection. Show that D_8/Z is generated by $p_2(a)$ and $p_2(b)$ which are of order two (and, of course, commute). Hence $D_8/Z \approx \mathbb{Z}_2 \oplus \mathbb{Z}_2$.

What we want to do with D_8 is to construct a sequence of C_i's like (27) and a sequence of A_i's like (28) and classes $\alpha_1 \in H^2(C_1; A_1)^{C_2}$ and $\alpha_2 \in H^2(C_2; A_2)$ with $\operatorname{res}(\alpha_2) = (i_A)_*(\alpha_1)$, but with the corresponding subgroup B_1 not normal in B_2. The exact sequence going across the middle

$$0 \longrightarrow A_2 \longrightarrow B_2 \longrightarrow C_2 \longrightarrow 0$$

will be

$$0 \longrightarrow Z \longrightarrow D_8 \longrightarrow \mathbb{Z}_2 \oplus \mathbb{Z}_2 \longrightarrow 0, \tag{32}$$

i.e. $A_2 = Z, B_2 = D_8$, and $C_2 = D_8/Z \approx \mathbb{Z}_2 \oplus \mathbb{Z}_2$.

C_1 will be the subgroup of C_2 generated by $p_2(b)$, so $C_2 \approx \mathbb{Z}_2$. This forces $C_3 = C_2/C_1 \approx \mathbb{Z}_2$ also.

The trick which makes everything easy to compute is to take $A_1 = 0$, so $A_3 = A_2 = Z \approx \mathbb{Z}_2$. Now (32) gives an action of C_2 on A_2, but since

$A_2 \approx \mathbb{Z}_2$ which has a trivial automorphism group, this action is trivial, and therefore the action of C_1 on A_2 is also trivial.

Now we define a section for the extension (32). The obvious thing to do is to set $s(p_2(a)) = a$, $s(p_2(b)) = b$, and $s(p_2(a)p_2(b)) = ab$. Let $c : C_2 \times C_2 \to Z$ be the associated 2-cocycle, i.e.

$$c(x, y) = s(x)s(y)s(xy)^{-1}.$$

Let $\alpha_2 \in H^2(C_2; Z)$ be the cohomology class of c and hence of the extension (32). We have $c(p_2(b), p_2(b)) = b^2 = 1$, and this is the only possibly non-trivial value of c on C_1 ($\approx \mathbb{Z}_2$), so $c|C_1$ is trivial, and $\text{res}(\alpha_2)$ is trivial.

Notice that $H^2(C_1; A_1)$ is 0 because $A_1 = 0$. Thus the only possibility for α_1 is 0. The top row is, of course,

$$0 \longrightarrow 0 \longrightarrow B_1 \longrightarrow C_1 \longrightarrow 0$$

where B_1 is the subgroup of D_8 generated by b. B_1 and C_1 are both isomorphic to $\mathbb{Z}_2$. Clearly $\alpha_1 \in H^2(C_1; A_1)^{C_2}$ since everything is 0. To finish the example, we note that B_1 is not normal in B_2 even though $\text{res}(\alpha_2) = \alpha_1$ $(= 0)$. We summarize in the following diagram:

$$\begin{array}{ccccccccc}
 & & 0 & & 0 & & 0 & & \\
 & & \downarrow & & \downarrow & & \downarrow & & \\
0 & \longrightarrow & 0 & \longrightarrow & B_1 = \langle b \rangle & \longrightarrow & C_1 = \langle p_2(b) \rangle & \longrightarrow & 0 \\
 & & \downarrow & & \downarrow \text{not normal} & & \downarrow & & \\
0 & \longrightarrow & Z = \langle a \rangle & \longrightarrow & B_2 = D_8 & \longrightarrow & C_2 = \langle p_2(a), p_2(b) \rangle & \longrightarrow & 0 \\
 & & \downarrow & & \downarrow & & \downarrow & & \\
0 & \longrightarrow & \mathbb{Z}_2 & \longrightarrow & \text{doesn't exist} & \longrightarrow & C_3 = \langle p_2(a) \rangle & \longrightarrow & 0 \\
 & & \downarrow & & & & \downarrow & & \\
 & & 0 & & & & 0 & &
\end{array}$$

and this concludes the example.

The following definitions will enable us to talk about the kinds of cochains and cohomology classes that arise.

Definition 3.2: Suppose we have a pair of suitable exact sequences (i.e. sequences (27)and (28) with A_3 a trivial C_1-module). A cochain $c \in C^q(C_2; A_2)$ is said to be *depressive* if

$$c(\gamma_1\delta_1, \gamma_2\delta_2, \ldots, \gamma_q\delta_q) \equiv c(\delta_1, \delta_2, \ldots, \delta_q) \ (\mathrm{mod} A_1) \tag{33}$$

for $\gamma_i \in C_1$ and $\delta_i \in C_2$. A cohomology class $\alpha \in H^q(C_2; A_2)$ is said to be *depressive* if it contains a depressive cocycle.

Exercise 3.5: i) Show that the set of depressive cochains in $C^q(C_2; A_2)$ forms a subgroup.

ii) Show that $H^q_{\mathrm{dep}}(C_2; A_2) = (p_A)_*^{-1}\left[p_C^*(H^q(C_3; A_3))\right]$.

We denote the subset of depressive cochains by $C^q_{\mathrm{dep}}(C_2; A_2)$ and the subgroup of depressive classes by $H^q_{\mathrm{dep}}(C_2; A_2)$. We define the *repression* map

$$\mathrm{rep} : C^q_{\mathrm{dep}}(C_2; A_2) \to C^q(C_1; A_1)$$

to be the restriction to C_1. This takes a little explanation. Suppose $c \in C^q_{\mathrm{dep}}(C_2; A_2)$. Then for $\gamma_i \in C_1$, we have

$$\begin{aligned} p_A(c(\gamma_1, \ldots, \gamma_q)) &= p_A(c(\gamma_1 \cdot 1, \ldots, \gamma_q \cdot 1)) \\ &= p_A(c(1, \ldots, 1)) \\ &= 1 \end{aligned}$$

by (33) and the fact that c is normalized (as are all of our cochains). Thus the restriction of a depressive cochain to C_1 *does* take values in A_1.

Similarly we have a map

$$\mathrm{dep} : C^q_{\mathrm{dep}}(C_2; A_2) \to C^q(C_3; A_3)$$

defined by

$$[\mathrm{dep}(c)]\,(\beta_1, \ldots, \beta_q) = p_A \circ c(\bar{\beta}_1, \ldots, \bar{\beta}_q)$$

where $c \in C^q_{\mathrm{dep}}(C_2; A_2)$, $\beta_i \in C_3$, and $\bar{\beta}_i \in C_2$ with $p_C(\bar{\beta}_i) = \beta_i$. You can see that $\mathrm{dep}(c)$ is well-defined by using (33). We call dep the *depression map*.

Lemma 3.3: *i) Rep is a homomorphism.*

ii) Dep is a homomorphism.

iii) Let $Z^q_{\text{dep}}(C_2; A_2)$ be the subgroup of depressive cocycles. Then

$$\text{dep}\left(Z^q_{\text{dep}}(C_2; A_2))\right) \subset Z^q(C_3; A_3),$$

and

$$\text{rep}\left(Z^q_{\text{dep}}(C_2; A_2))\right) \subset Z^q(C_1; A_1).$$

iv) A class $z \in Z^q(C_2; A_2)$ is depressive iff

$$z(\delta_1, \ldots, \delta_{i-1}, \gamma, \delta_{i+1}, \ldots, \delta_q) \in A_1 \tag{34}$$

whenever $\delta_j \in C_2$ and $\gamma \in C_1$.

v) The sequence

$$0 \longrightarrow Z^q(C_2; A_1) \xrightarrow{i_A} Z^q_{\text{dep}}(C_2; A_2) \xrightarrow{\text{dep}} Z^q(C_3; A_3)$$

is exact.

vi) The sequence

$$0 \longrightarrow Z^1(C_3; A_2) \xrightarrow{\text{inf}} Z^1_{\text{dep}}(C_2; A_2) \xrightarrow{\text{rep}} Z^1(C_1; A_1)$$

is exact.

Proof: i), ii), iii) are easy and will be left as exercises. For iv), suppose that z satisfies (34). We have, by Equation (16) of Chapter III,

$$\begin{aligned} 0 = \delta z(\delta_1, \ldots, \delta_{i-1}, \gamma, \delta_i, \delta_{i+1}, \ldots, \delta_q) &= \delta_1 \cdot z(\delta_2, \ldots, \delta_{i-1}, \gamma, \delta_i, \ldots, \delta_q) \\ &+ \sum_{j=1}^{q} (-1)^j z(\delta_1, \ldots, \delta_{i-1}, \gamma, \delta_i, \ldots, \delta_j \delta_{j+1}, \ldots, \delta_q) \\ &+ (-1)^{q+1} z(\delta_1, \ldots, \delta_{i-1}, \gamma, \delta_i, \ldots, \delta_{q-1}) \end{aligned}$$

Now take p_A of both sides. Recall that by (34), $p_A(z)$ is 0 if any of its arguments lie in C_1. Suppose $i = 1$. Then we get

$$[p_A(z)](\delta_1, \ldots, \delta_q) = [p_A(z)](\gamma\delta_1, \ldots, \delta_q)$$

since all the other terms are 0 and C_1 acts trivially on A_3. Suppose $1 < i \le q$. Then we get

$$[p_A(z)](\delta_1, \ldots, \delta_{i-1}\gamma, \delta_i, \ldots, \delta_q) = [p_A(z)](\delta_1, \ldots, \delta_{i-1}, \gamma\delta_i, \ldots, \delta_q).$$

Suppose $i = q+1$. Then we get

$$[p_A(z)](\delta_1, \ldots, \delta_q\gamma) = [p_A(z)](\delta_1, \ldots, \delta_q).$$

Observe that we have

$$[p_A(z)](\ldots, \delta_{i-1}\gamma, \ldots) = [p_A(z)](\ldots, \gamma\delta_i, \ldots)$$

for all $\gamma \in C_1$. We have

$$[p_A(z)](\delta_1, \ldots, \delta_q) = [p_A(z)](\gamma\delta_1, \ldots, \delta_q),$$

and letting $\gamma' = \delta_1\gamma\delta_1^{-1}$, since C_1 is normal in C_2, we get

$$\begin{aligned}[p_A(z)](\delta_1, \ldots, \delta_q) &= [p_A(z)](\gamma'\delta_1, \ldots, \delta_q)\\ &= [p_A(z)](\delta_1\gamma, \delta_2, \ldots, \delta_q)\\ &= [p_A(z)](\delta_1, \gamma\delta_2, \ldots, \delta_q),\end{aligned}$$

and letting $\gamma'' = \delta_2\gamma\delta_2^{-1}$, since C_1 is normal in C_2, we get

$$\begin{aligned}[p_A(z)](\delta_1, \ldots, \delta_q) &= [p_A(z)](\delta_1, \gamma''\delta_2, \ldots, \delta_q)\\ &= [p_A(z)](\delta_1, \delta_2, \gamma\delta_3, \ldots, \delta_q).\end{aligned}$$

Continuing, we get

$$[p_A(z)](\delta_1, \ldots, \delta_q) = [p_A(z)](\delta_1, \ldots, \delta_{i-1}, \gamma\delta_i, \ldots, \delta_q),$$

for $1 \leq i \leq q$. Hence

$$\begin{aligned}[p_A(z)](\gamma_1\delta_1, \gamma_2\delta_2, \ldots, \gamma_q\delta_q) &= [p_A(z)](\delta_1, \gamma_2\delta_2, \ldots, \gamma_q\delta_q)\\ &= [p_A(z)](\delta_1, \delta_2, \gamma_3\delta_3, \ldots, \gamma_q\delta_q)\\ &= \ldots\\ &= [p_A(z)](\delta_1, \delta_2, \ldots, \delta_q)\end{aligned}$$

which is precisely (33) so c is depressive.

Conversely if z is depressive (i.e. satisfies (33)), then as above we have

$$\begin{aligned}[p_A(z)](\delta_1, \ldots, \delta_{i-1}, \gamma, \delta_{i+1}, \ldots, \delta_q) &= [p_A(z)](\delta_1, \ldots, \gamma\cdot 1, \ldots, \delta_q)\\ &= [p_A(z)](\delta_1, \ldots, \delta_{i-1}, 1, \delta_{i+1}, \ldots, \delta_q)\\ &= 0,\end{aligned}$$

so c satisfies (34).

For v), it is clear from (34) that i_A maps $Z^q(C_2; A_1)$ into $Z^q_{\text{dep}}(C_2; A_2)$ since anything in its image takes all of its values in A_1. Since i_A is injective, it is injective on cocycles. Furthermore dep involves projection via p_A, so $\text{dep} \circ i_A = 0$. Finally suppose $\text{dep}(z) = 0$, i.e. $z \in Z^q_{\text{dep}}(C_2; A_2)$, and

$$[\text{dep}(z)]\,(\beta_1, \ldots, \beta_q) = p_A \circ z(\bar{\beta}_1, \ldots, \bar{\beta}_q) = 0 \tag{35}$$

where $\beta_i \in C_3$, and $\bar{\beta}_i \in C_2$ with $p_C(\bar{\beta}_i) = \beta_i$. Define $z' \in C^q(C_2; A_1)$ by

$$z'(\delta_1, \ldots, \delta_q) = {i_A}^{-1} \circ z(\delta_1, \ldots, \delta_q). \tag{36}$$

To see that (36) makes sense we must show that $z(\delta_1, \ldots, \delta_q) \in \text{img} i_A$, but that is what (35) says. It is clear that $i_A(z') = z$, so it suffices to show that z' is a cocycle which is straightforward and omitted.

For vi), recall that

$$[\inf(z)](\delta) = z(p_C(\delta))$$

for $z \in Z^1(C_3; A_2)$, and $\delta \in C_2$, so it is clear that inf takes its image in $Z^1_{\text{dep}}(C_2; A_2)$, and also that inf is injective since p_C is surjective. It is obvious that $\text{rep} \circ \inf = 0$. Now suppose that $\text{rep}(z) = 0$, i.e. $z|C_1$ is 0. Define $z' \in C^1(C_3; A_2)$ by

$$z'(\beta) = z(\bar{\beta})$$

where $p_C(\bar{\beta}) = \beta$. This is well-defined because if $\gamma \in C_1$,

$$z(\bar{\beta}\gamma) = \bar{\beta} \cdot z(\gamma) + z(\bar{\beta}) = z(\bar{\beta}) \tag{37}$$

since z is a 1-cocycle, and $z|C_1 = 0$. Clearly $\inf(z') = z$ and z' is a cocycle. □

Remark: One might be tempted by iii) to suppose that dep and rep define maps on cohomology. The problem is that, in general, dep and rep do not carry coboundaries into coboundaries. In addition, v) might tempt one into trying to prove vi) for $q > 1$ (i.e. start with $Z^q(C_3; A_3)$), but if you examine the proof of vi), you will see in equation (37) a crucial use of the 1-cocycle formula. The 2-cocycle formula will not work here.

The next theorem is the basic existence theorem.

Theorem 3.1 : *Suppose we have a short exact sequence of groups*

$$1 \longrightarrow C_1 \xrightarrow{i_C} C_2 \xrightarrow{p_C} C_3 \longrightarrow 1,$$

and a short exact sequence of C_2-modules

$$0 \longrightarrow A_1 \xrightarrow{i_A} A_2 \xrightarrow{p_A} A_3 \longrightarrow 0,$$

with A_3 a trivial C_1-module.

i) If

$$1 \longrightarrow B_1 \xrightarrow{i_B} B_2 \xrightarrow{p_B} B_3 \longrightarrow 1,$$

is a short exact sequence of groups which fits into the following 9-diagram:

$$\begin{array}{ccccccccc}
 & & 0 & & 1 & & 1 & & \\
 & & \downarrow & & \downarrow & & \downarrow & & \\
0 & \longrightarrow & A_1 & \xrightarrow{i_1} & B_1 & \xrightarrow{p_1} & C_1 & \longrightarrow & 1 \\
 & & \downarrow{\scriptstyle i_A} & & \downarrow{\scriptstyle i_B} & & \downarrow{\scriptstyle i_C} & & \\
0 & \longrightarrow & A_2 & \xrightarrow{i_2} & B_2 & \xrightarrow{p_2} & C_2 & \longrightarrow & 1 \\
 & & \downarrow{\scriptstyle p_A} & & \downarrow{\scriptstyle p_B} & & \downarrow{\scriptstyle p_C} & & \\
0 & \longrightarrow & A_3 & \xrightarrow{i_3} & B_3 & \xrightarrow{p_3} & C_3 & \longrightarrow & 1 \\
 & & \downarrow & & \downarrow & & \downarrow & & \\
 & & 0 & & 1 & & 1, & &
\end{array}$$

then there is a depressive 2-cocycle $z_2 \in Z^2_{\text{dep}}(C_2; A_2)$ s.t. the cohomology class of the extension

$$0 \longrightarrow A_j \xrightarrow{i_j} B_j \xrightarrow{p_j} C_j \longrightarrow 1$$

is the class of z_j $(j = 1, 2, 3)$ and $z_1 = \text{rep}(z_2)$ and $z_3 = \text{dep}(z_2)$.

ii) Suppose $z_2 \in Z^2_{\text{dep}}(C_2; A_2)$. Put $z_1 = \text{rep}(z_2)$ and $z_3 = \text{dep}(z_2)$. Then we can choose an extension B_j associated to the class of z_j (j=1,2,3) s.t.

$$1 \longrightarrow B_1 \xrightarrow{i_B} B_2 \xrightarrow{p_B} B_3 \longrightarrow 1$$

is exact and fits into the above 9-diagram.

Proof: i) is essentially Proposition 3.1 except now we must show that α_2 (the class of z_2) is depressive. But α_2 satisfies $\text{res}(\alpha_2) = (i_A)_*(\alpha_1)$ and part v) of Lemma 3.3 shows that the image of $(i_A)_*$ is in $Z^2_{\text{dep}}(C_2; A_2)$.

For ii) let B_2 be the extension constructed with the 2-cocycle z_2, i.e. $B_2 = C_2 \times A_2$ as a set and

$$(c, a)(c', a') = (cc', c \cdot a' + a + z_2(c, c'))$$

where we are using additive notation in A_2. Let B_1 be the subset of B_2 given by

$$B_1 = \{(c, a) \in B_2 : c \in C_1 \text{ and } a \in A_1\}.$$

Equation (34) and the fact that z_2 is depressive show that B_1 is a subgroup of B_2. Furthermore B_1 is an extension of A_1 by C_1 associated to the 2-cocycle $z_1 = \text{rep}(z_2)$.

Claim: B_1 is normal in B_2.

Let $c \in C_2, a \in A_2, c_1 \in C_1$, and $a_1 \in A_1$. Then

$$(c, a)(c_1, a_1)(c, a)^{-1} = (cc_1c^{-1}, (1 - cc_1c^{-1}) \cdot a + \\ c \cdot a_1 + z_2(c, c_1) - cc_1c^{-1} \cdot z_2(c, c^{-1}) + z_2(cc_1, c^{-1})).$$

Since C_1 is normal in C_2, $c' = cc_1c^{-1}$ is in C_1. Since C_1 acts trivially on A_3, $(1 - c') \cdot a \in A_1$, i.e. $p_A(a - c' \cdot a) = 0$. Clearly $c \cdot a_1 \in A_1$ and $z_2(c, c_1) \in A_1$ since $c_1 \in C_1$ and z_2 is depressive. Now since $\delta z_2(c, c_1, c^{-1}) = 0$, we get

$$z_2(cc_1, c^{-1}) = c \cdot z_2(c_1, c^{-1}) + z_2(c, c_1c^{-1}) - z_2(c, c_1) = z_2(c, c^{-1}) + a'$$

for some $a' \in A_1$ (recall that $z_2(c, c_1c^{-1}) \equiv z_2(c, c^{-1}) \ (\text{mod} A_1)$). Hence the last two terms are $(1 - c') \cdot z_2(c, c^{-1})+$ something in A_1, and, as above, $(1 - c') \cdot z_2(c, c^{-1})$ is in A_1 also. Hence $(c, a)(c_1, a_1)(c, a)^{-1} \in B_1$, and B_1 is normal in B_2.

Let $B_3 = B_2/B_1$.

Exercise 3.6: i) Show that B_3 is an extension of A_3 by C_3 which corresponds to the cohomology class of $z_3 = \text{dep}(z_2)$.

ii) Show that the exact sequence

$$1 \longrightarrow B_1 \xrightarrow{i_B} B_2 \xrightarrow{p_B} B_3 \longrightarrow 1,$$

fits into the 9-diagram. (**Hint:** Since all of the maps are already defined, all you have to check is the commutativity of the diagram.)

This concludes the proof of the theorem. □

Now we want to ask about uniqueness, i.e. what happens if we change the depressive 2-cocycle z_2 by a 2-coboundary? In terms of the extension B_2, this change just amounts to using a different section : $C_2 \to B_2$. We want to know how it affects $\operatorname{dep}(z_2)$ and $\operatorname{rep}(z_2)$. In order to do this we have to restrict to changing z_2 by a depressive coboundary. since the depression or repression of a coboundary may not be a coboundary.

Recall the connection (or transgression) map from the last section,

$$\operatorname{con}_{C_3}^{C_1} : H^1(C_1; A_3)^{C_2} \to H^2(C_3; A_3).$$

Recall also the connecting homomorphism (or Bockstein) β from Theorem 3.2 of Chapter III. Since

$$0 \longrightarrow A_1 \xrightarrow{i_A} A_2 \xrightarrow{p_A} A_3 \longrightarrow 0$$

is an exact sequence of C_2-modules, we get a map

$$\beta^1 = \beta : H^1(C_1; A_3) \to H^2(C_1; A_1).$$

We use $B^i_{\text{dep}}(C_j; A_k)$ to be the obvious thing , the appropriate group of depressive coboundaries.

Well, now we have an abundance of maps. We have δ, and con, and β, and res, and inf, and rep, and dep. The next lemma tells us how some of them relate to others.

Lemma 3.4: *Using all of the preceeding notation, let $c \in C^1(C_2; A_2)$ satisfy $\delta c \in B^2_{\text{dep}}(C_2; A_2)$.*

i) Let $C^1_t(C_2; A_3)$ be the set of $g \in C^1(C_2; A_3)$ s.t.

a) $\operatorname{res}(g) \in Z^1(C_1; A_3)$ and

b) $\delta g \in \inf Z^2(C_3; A_3) \subset Z^2(C_2; A_3)$.

Then the following sequence is exact:

$$0 \longrightarrow i_A\left[C^1(C_2; A_1)\right] \cap \delta^{-1}\left[B^2_{\text{dep}}(C_2; A_2)\right] \longrightarrow \delta^{-1}[B^2_{\text{dep}}(C_2; A_2)] \xrightarrow{p_A} C^1_t(C_2; A_3) \longrightarrow 0.$$

We had better say this out in words. If you take $c \in C^1(C_2; A_2)$ with δc a depressive 2-coboundary and project it via p_A to $C^1(C_2; A_3)$, you get an element in $C^1_t(C_2; A_3)$, i.e. satisfying a) and b). Furthermore the projection of all such c's has as kernel, those c's which come from $C^1(C_2; A_1)$ via i_A.

ii) Let f be the map that sends c into the cohomology class of $\text{res} \circ p_A(c) = p_A \circ \text{res}(c)$*. Then the following sequence is exact:*

$$0 \longrightarrow C^1_{\text{dep}}(C_2; A_2) \longrightarrow \delta^{-1}\left[B^2_{\text{dep}}(C_2; A_2)\right] \xrightarrow{f} H^1(C_1; A_3)^{C_2} \longrightarrow 0.$$

iii) $\text{con}[\text{res} \circ p_A(c)] = [\text{dep}(\delta c)]$ *where "[ ]" denotes cohomology class.*

iv) $\beta[\text{res} \circ p_A(c)] = [\delta(\text{res} \circ p_A(c))] = [\text{rep}(\delta c)]$.

Proof: These statements are complicated but not at all deep. Let

$$\mathcal{B} = \delta^{-1}\left[B^2_{\text{dep}}(C_2; A_2)\right] \subset C^1(C_2; A_2),$$

i.e. $\mathcal{B}$ is where the c's in the statement of the lemma live.

For i) we first show that $p_A|\mathcal{B}$ takes its image in $C^1_t(C_2; A_3)$ So we take $c \in \mathcal{B}$. We want to show that $p_A(c)$ satisfies a) and b). For a) we want to show that $\delta \circ \text{res} \circ p_A(c) = 0$. But $\delta \circ \text{res} \circ p_A(c) = p_A \circ \text{res}(\delta c)$, and δc is depressive, so its restriction to C_1 takes values in A_1. Hence when we project to A_3, we get 0. For b) we want to show that there is $z \in Z^2(C_3; A_2)$ s.t. $\delta c = \inf z$. Since p_C is surjective, inf is injective, so we can take $z = \inf^{-1}(\delta c)$ which clearly lies in $Z^2(C_3; A_2)$.

To see that $p_A|\mathcal{B}$ is surjective, let $g \in C^1_t(C_2; A_3)$. We want $c \in \mathcal{B}$ with $p_A(c) = 0$. Notice that

$$0 \longrightarrow C^1(C_2; A_1) \xrightarrow{i_A} C^1(C_2; A_2) \xrightarrow{p_A} C^1(C_2; A_3) \longrightarrow 0 \tag{38}$$

is exact. Let $\bar{g} \in p_A^{-1}(g) \subset C^1(C_2; A_2)$. Clearly $p_A(\bar{g}) = g$, so if we can show that $\delta\bar{g}$ is depressive, we will have $c = \bar{g}$. Let $\gamma_1 \in C_1$, and $\gamma_2 \in C_2$, and compute

$$\begin{aligned} p_A \circ \delta \circ \bar{g}(\gamma_1, \gamma_2) &= \delta g(\gamma_1, \gamma_2) \\ &= \inf z(\gamma_1, \gamma_2) \\ &\qquad \text{for some } z \in Z^2(C_3; A_3) \text{ by b)} \\ &= z(p_C(\gamma_1), p_C(\gamma_2)) \\ &= z(0, p_C(\gamma_2)) \\ &= 0, \end{aligned}$$

so $\delta\bar{g}(\gamma_1,\gamma_2)$ is in A_1, and $\delta\bar{g}$ is depressive. Now Sequence (38) shows the rest of the exactness quite easily.

For ii) let's show that f takes values in $H^1(C_1;A_3)^{C_2}$. By i), $p_A(c) \in C_t^1(C_2;A_3)$, $\mathrm{res}\circ p_A(c) \in Z^1(C_1;A_3)$, and f is indeed well-defined as a map to $H^1(C_1;A_3)$. By Lemma 2.1, its image is in $H^1(C_1;A_3)^{C_2}$ because the image of res is always there.

Now the restriction of a depressive 1-cochain takes values in A_1, so when it is projected by p_A, we get 0. Hence $C^1_{\mathrm{dep}}(C_2;A_2) \subset \ker f$. On the other hand, suppose we have $c \in \mathcal{B}$ and $\mathrm{res}\circ p_A(c)$ is cohomologous to 0, i.e. there is $b \in C^0(C_1;A_3)$ with $\mathrm{res}\circ p_A(c) = \delta b$. Since C_1 acts trivially on A_3, $\delta b = 0$, so the projection by p_A of the restriction of c is 0, i.e. c takes elements in C_1 to A_1, i.e. c is depressive.

We have, in fact, already done the necessary computation for iii). Part iii) says that

$$p_A \circ c(a(\gamma,\gamma')) = p_A\delta c(s(\gamma), s(\gamma')),$$

where $\gamma,\gamma' \in C_3$ and $a(\gamma,\gamma') = s(\gamma)s(\gamma')s(\gamma\gamma')^{-1}$ as usual. This is precisely the computation following Equation (21) with $K = C_1$, $A = A_3$, $c = \mathrm{res}\circ p_A(c)$, and the b there is the same as the c there since we are evaluating them on elements of C_1.

Part iv) is essentially just definitions. Recall the definition of the Bockstein β. If $z \in Z^1(C_1;A_3)$, let $\bar{z} \in p_A^{-1}(z) \subset C^1(C_2;A_3)$. Then $\beta(z) = \delta\bar{z}$, and it is easy to see this is in $Z^2(C_1;A_1)$, and that β is well-defined on cohomology. Hence

$$\beta\left[\mathrm{res}\circ p_A(c)\right] = \left[\delta\circ p_A^{-1}\circ \mathrm{res} p_A(c)\right] = \left[\mathrm{res}(\delta c)\right]$$

which is $[\mathrm{rep}(\delta c)]$ by the definition of rep. □

We now use this lemma to see how the class of $\mathrm{dep}(z)$ and $\mathrm{rep}(z)$ (for z a depressive 2-cocycle) change when we alter z within its cohomology class.

Theorem 3.2 : *Suppose we have a short exact sequence of groups*

$$1 \longrightarrow C_1 \xrightarrow{i_C} C_2 \xrightarrow{p_C} C_3 \longrightarrow 1,$$

and a short exact sequence of C_2-modules

$$0 \longrightarrow A_1 \xrightarrow{i_A} A_2 \xrightarrow{p_A} A_3 \longrightarrow 0,$$

with A_3 a trivial C_1-module.

i) Let z and z' be depressive 2-cocycles in $Z^2_{\mathrm{dep}}(C_2; A_2)$ in the same cohomology class. Then there is a cohomology class $[u] \in H^1(C_1; A_3)^{C_2}$ s.t.

$$a)\ [\mathrm{dep}(z')] = [\mathrm{dep}(z)] + \mathrm{con}([u]) \text{ and}$$
$$b)\ [\mathrm{rep}(z')] = [\mathrm{rep}(z)] + \beta([u]).$$

ii) Suppose $z \in Z^2_{\mathrm{dep}}(C_2; A_2)$ and $[u] \in H^1(C_1; A_3)^{C_2}$ are given. Then there is a 2-cocycle $z' \in Z^2_{\mathrm{dep}}(C_2; A_2)$ s.t. z and z' are cohomologous and s.t. a) and b) hold.

Proof: We show that i) is an easy consequence of the lemma. If z and z' are cohomologous, then $z - z' = \delta c$ for some $c \in C^1(C_2; A_2)$, and δc depressive says that for $c \in \mathcal{B}$, Lemma 3.4 applies to c. By ii) of that lemma, $[\mathrm{res} \circ p_A(c)] \in H^1(C_1; A_3)^{C_2}$, so let $u = \mathrm{res} \circ p_A(c)$. Then a) is simply iii) of the lemma and b) is part iv).

Part ii) of our theorem is similar. By ii) of Lemma 3.4, there is $c \in \mathcal{B}$ s.t. $\mathrm{res} \circ p_A(c) = u$. Then we can put $z' = z + \delta c$. $\square$

Remark: Let's review what progress we have on the problem of "filling in" the middle column of a 9-diagram. Proposition 3.1 gave us necessary conditions on the cohomology classes of the group extensions which would, we hope, be the rows of the completed 9-diagram, but our Example 3.1 showed that these cohomological conditions were not sufficient. We then had to look at a particular type of 2-cocycle to get better results. Theorem 3.1 tells us that a cohomology class in $H^2(C_2; A_2)$ which contains such a depressive 2-cocycle does give rise to a "filling in" of the 9-diagram, in fact, many different "fillings in", i.e. many different choices of B_1 and B_3 can be obtained to go with the B_2 determined by the depressive class. Just take any classes γ_1 and α_2 determining B_1 and B_2 respectively, and then take any class $\gamma \in H^1(C_1; A_3)^{C_2}$, and then add $\beta(\gamma)$ and $\mathrm{con}(\gamma)$ to the classes α_1 and α_2 respectively.

To get a handle on this situation, we would like to say that each fixed depressive $\alpha_2 \in H^2_{\mathrm{dep}}(C_2; A_2)$ determines a map between admissible classes $\alpha_1 \in H^2(C_1; A_1)^{C_2}$ (recall that α_1 must be left fixed by C_2 because of Proposition 3.1) and $\alpha_3 \in H^2(C_3; A_3)$ as follows: Pick a depressive $z_2 \in \alpha_2$. Then pick any $\alpha \in H^1(C_1; A_3)^{C_2}$ s.t. $[\mathrm{rep}(z_2)] - \alpha_2 = \beta(\alpha)$. Then we would like to set $\alpha_3 = [\mathrm{dep}(z_2)] + \mathrm{con}(\alpha)$. A glance at Equations

a) and b) of Theorem 3.2 shows that this works with $[u] = \gamma$ and with z_2' equal to the depressive 2-cocycle with $[\mathrm{dep}(z_2')] = \alpha_3$ and $[\mathrm{rep}(z_2')] = \alpha_1$. The trouble with this is that, in general, γ will *not* be unique.

Exercise 3.7: Show that β takes $H^1(C_1; A_3)^{C_2}$ into $H^2(C_1; A_1)^{C_2}$. (**Hint:** Recall $\beta(\gamma) = [\delta p_A^{-1}(z)]$ for $z \in \gamma$, and compute $\lambda * \beta(\gamma)$ for $\lambda \in C_2$.)

Now we have the simple diagram

$$\begin{array}{ccc} H^1(C_1; A_3)^{C_2} & \xrightarrow{\text{con}} & H^2(C_3; A_3) \\ \big\downarrow{\scriptstyle\beta} & & \\ H^2(C_1; A_1)^{C_2}. & & \end{array}$$

It is easy to see that we get a homomorphism from $\operatorname{img}\beta$ to $\operatorname{img}\mathrm{con}$ iff $\ker\beta \subset \ker\mathrm{con}$. The Hochschild–Serre exact sequence (Theorem 2.1) tells us that $\ker\mathrm{con}$ is $\mathrm{res}(H^1(C_2; A_3))$. The cohomology exact sequence (Theorem 3.2 of Chapter III) tells us that $\ker\beta$ equals $p_A(H^1(C_1; A_2))^{C_2}$. (We must be careful here because $\left(p_A(H^1(C_1; A_2))\right)^{C_2}$ may not be equal to $p_A\left(H^1(C_1; A_2)^{C_2}\right)$.) Summarizing, we have the following:

Proposition 3.2: *In the situation of Theorem 3.2, if*

$$\left(p_A(H^1(C_1; A_2))\right)^{C_2} \subset \mathrm{res}(H^1(C_2; A_3)),$$

then each depressive cohomology class $\alpha_2 \in H^2_{\mathrm{dep}}(C_2; A_2)$ determines a homomorphism between $H^2(C_1; A_1)^{C_2}$ and $H^2(C_3; A_3)$ which takes a class α_1 corresponding to a subgroup B_1 of B_2 (the group corresponding to α_2) into the class of the group $B_3 = B_2/B_1$. This homomorphism is defined by the formulas

$$\alpha_3 = [\mathrm{dep}(z_2)] + \mathrm{con}(\gamma)$$

where z_2 is any depressive 2-cocycle in α_2, and γ is chosen in $H^1(C_1; A_3)^{C_2}$ s.t.

$$\beta(\gamma) = [\mathrm{rep}(z_2)] - \alpha_1.$$

The last theorem of this section concerns automorphisms and gives another interpretation of $Z^2_{\mathrm{dep}}(C_2; A_2)$.

Definition 3.3: Suppose we have a 9-diagram

$$\begin{array}{ccccccccc}
 & & 0 & & 1 & & 1 & & \\
 & & \downarrow & & \downarrow & & \downarrow & & \\
0 & \longrightarrow & A_1 & \xrightarrow{i_1} & B_1 & \xrightarrow{p_1} & C_1 & \longrightarrow & 1 \\
 & & \downarrow{\scriptstyle i_A} & & \downarrow{\scriptstyle i_B} & & \downarrow{\scriptstyle i_C} & & \\
0 & \longrightarrow & A_2 & \xrightarrow{i_2} & B_2 & \xrightarrow{p_2} & C_2 & \longrightarrow & 1 \\
 & & \downarrow{\scriptstyle p_A} & & \downarrow{\scriptstyle p_B} & & \downarrow{\scriptstyle p_C} & & \\
0 & \longrightarrow & A_3 & \xrightarrow{i_3} & B_3 & \xrightarrow{p_3} & C_3 & \longrightarrow & 1 \\
 & & \downarrow & & \downarrow & & \downarrow & & \\
 & & 0 & & 1 & & 1. & &
\end{array}$$

As before we use $\mathrm{Aut}^0(B_2)$ to denote the group of those automorphisms of B_2 which induce the identity on A_2 and C_2. An *automorphism of the* 9-*diagram* will be an element of $\mathrm{Aut}^0(B_2)$ which carries B_1 into B_1 (and hence induces an automorphism of B_3). This subgroup of $\mathrm{Aut}^0(B_2)$ will be denoted by $\mathrm{Aut}^9(B_2)$.

Recall that $Z^1(C_2; A_2)$ is isomorphic to $\mathrm{Aut}^0(B_2)$ and $B^1(C_2; A_3)$ is isomorphic to $\mathrm{Aut}(B_2) \cap \mathrm{Inn}(B_2)$ (see Lemma 1.2).

Theorem 3.3 : *Assume we have a 9-diagram as above.*

i) The group $Z^1_{\mathrm{dep}}(C_2; A_2)$ *is isomorphic to* $\mathrm{Aut}^9(B_2)$ *and furthermore,* $B^1(C_2; A_2)$ *is a subgroup of* $Z^1_{\mathrm{dep}}(C_2; A_2)$.

ii) If $z \in Z^1_{\mathrm{dep}}(C_2; A_2)$, *then the automorphism of the* 9-*diagram corresponding to* z *induces on* B_1 *the automorphism corresponding to* $\mathrm{rep}(z)$, *and on* B_3 *the automorphism corresponding to* $\mathrm{dep}(z)$.

iii) The map dep induces a homomorphism

$$H^1_{\mathrm{dep}}(C_2; A_2) \to H^1(C_3; A_3)$$

whose kernel is $i_A(H^1(C_2; A_1))$.

iv) The map rep induces a homomorphism

$$Z^2_{\mathrm{dep}}(C_2; A_2) \to H^1(C_1; A_1)^{C_2}.$$

v) We have

$$i_A([\text{rep}(z)]) = \text{res}([z])$$

and

$$[\text{dep}(z)] = p_A([z]).$$

Proof: The first statement of i) follows from the proof of Lemma 1.2, i.e. if z is depressive, it is easy to see that the corresponding $\psi \in \text{Aut}^0(B_2)$ carries B_1 into B_1. The second statement of i) follows from the definitions, i.e. if $a_2 \in A_2$ is associated to $b \in B^1(C_2; A_2)$ (i.e. $b(c_2) = c_2 \cdot a_2 - a_2$), then $c_1 \in C_1$ implies $b(c_1) = c_1 \cdot a_2 - a_2$ which is in A_1 since C_1 acts trivially on A_3.

Exercise 3.8: Prove ii), iii), iv), and v).

This concludes the proof and the section. □

4. Automorphisms of Group Extensions

In this section we start with a group extension

$$0 \longrightarrow M \longrightarrow \pi \longrightarrow \Phi \longrightarrow 1 \tag{39}$$

in which we assume that M is abelian, but that's almost all. We want to study $\text{Aut}(\pi, M)$, the group of those automorphisms of π that carry M into M. If π is Bieberbach, then, of course, $\text{Aut}(\pi, M) = \text{Aut}(\pi)$. We make one additional assumption. We want Lemma 1.3 to be true, or, equivalently, we want the top row of the basic diagram,

$$0 \longrightarrow M/M^\Phi \longrightarrow \text{Inn}(\pi) \longrightarrow \Phi \longrightarrow 1, \tag{40}$$

to be exact.

Exercise 4.1: i) Show that (40) is exact iff M^Φ is the center of π.

ii) Show that if M is a faithful Φ-module, M^Φ is the center of π.

Unless we say the contrary, we will assume in this section that M^Φ is the center of π. With this assumption we can construct a basic diagram like (6), but with $\text{Aut}(\pi)$ replaced by $\text{Aut}(\pi, M)$. We have $A_1 = M/M^\Phi, B_1 = \text{Inn}(\pi)$, and $C_1 = \Phi$. Let $\alpha_1 \in H^2(C_1; A_1) = H^2(\Phi; M/M^\Phi)$ be the cohomology class of the extension (40). As ususal, we let $\alpha \in H^2(\Phi; M)$ be the cohomology class of (39).

Exercise 4.2: Let $\rho : M \to M/M^{\Phi}$ be the projection. Show that $\alpha_1 = \rho_*(\alpha)$.

As before we let $A_2 = \mathrm{Aut}^0(\pi)$, the group of those automorphisms of π which induce the identity on M and Φ. As in Lemma 1.2, A_1 is isomorphic to $B^1(\Phi; M)$ and A_2 is isomorphic to $Z^1(\Phi; M)$, so $A_3 = A_2/A_1$ is isomorphic to $H^1(\Phi; M)$. We can let B_2 be any subgroup of $\mathrm{Aut}(\pi, M)$ which contains A_2 and B_1. We are, of course, interested in the case $B_2 = \mathrm{Aut}(\pi, M)$.

Exercise 4.3: If $B_2 = \mathrm{Aut}(\pi, M)$, show that $C_2 = B_2/A_2$ is isomorphic to the group of all semi-linear automorphismsm of M which fix α, i.e. if (f, A) is a semi-linear automorphism, then $f_*(\alpha) = A^*(\alpha)$. (**Hint:** This is Theorem 2.1 of Chapter III.)

In general, the identity map on M can be semi-linear w.r.t. many automorphisms of Φ, i.e. if a piece of Φ acts trivially on all of M, then the identity map is semi-linear w.r.t. any automorphism of that piece. If M is faithful, however, then the identity map on M is semi-linear w.r.t. only the identity map of Φ. In this case A_2 can be characterized as the group of automorphisms of π which induce the identity on M. We denote the group of semi-linear automorphisms of M by $\mathrm{Aut}_S(M)$ while the subgroup of those automorphisms which fix α will be denoted by $\mathrm{Aut}_S(M)_\alpha$.

Exercise 4.4: Show that if M is faithful, $\mathrm{Aut}_S(M)$ can be identified with a subgroup of $\mathrm{Aut}(M)$. If, in addition, M is free, then show that $\mathrm{Aut}_S(M)_\alpha = N_\alpha$. (**Hint:** Say $(f, A) \in \mathrm{Aut}_S(M)$. Show that, considered as an element of $\mathrm{Aut}(M)$, $A(\sigma) = f \circ \sigma \circ f^{-1}$.)

What is the goal of this section? The game is that we are given (39), so we know $\alpha \in H^2(\Phi; M)$. In addition, from the ideas of the first section we can construct a basic 9-diagram which we reproduce below together with the standard 9-diagram so you can match the various constituent elements.

$$\begin{array}{ccccccccc}
 & & 0 & & 1 & & 1 & & \\
 & & \downarrow & & \downarrow & & \downarrow & & \\
0 & \longrightarrow & M/M^{\Phi} & \longrightarrow & \mathrm{Inn}(\pi) & \longrightarrow & \Phi & \longrightarrow & 1 \\
 & & \downarrow & & \downarrow & & \downarrow & & \\
0 & \longrightarrow & \mathrm{Aut}^0(\pi) & \longrightarrow & \mathrm{Aut}(\pi, M) & \longrightarrow & \mathrm{Aut}_S(M)_\alpha & \longrightarrow & 1 \\
 & & \downarrow & & \downarrow & & \downarrow & & \\
0 & \longrightarrow & H^1(\Phi; M) & \longrightarrow & \mathrm{Out}(\pi, M) & \longrightarrow & \mathrm{Aut}_S(M)_\alpha/\Phi & \longrightarrow & 1 \\
 & & \downarrow & & \downarrow & & \downarrow & & \\
 & & 0 & & 1 & & 1 & &
\end{array}$$

$$\begin{array}{ccccccccc}
 & & 0 & & 1 & & 1 & & \\
 & & \downarrow & & \downarrow & & \downarrow & & \\
0 & \longrightarrow & A_1 & \xrightarrow{i_1} & B_1 & \xrightarrow{p_1} & C_1 & \longrightarrow & 1 \\
 & & \downarrow{\scriptstyle i_A} & & \downarrow{\scriptstyle i_B} & & \downarrow{\scriptstyle i_C} & & \\
0 & \longrightarrow & A_2 & \xrightarrow{i_2} & B_2 & \xrightarrow{p_2} & C_2 & \longrightarrow & 1 \\
 & & \downarrow{\scriptstyle p_A} & & \downarrow{\scriptstyle p_B} & & \downarrow{\scriptstyle p_C} & & \\
0 & \longrightarrow & A_3 & \xrightarrow{i_3} & B_3 & \xrightarrow{p_3} & C_3 & \longrightarrow & 1 \\
 & & \downarrow & & \downarrow & & \downarrow & & \\
 & & 0 & & 1 & & 1 & &
\end{array}$$

We want to compute $\alpha_1 \in H^2(C_1; A_1), \alpha_2 \in H^2(C_2; A_2)$, and $\alpha_3 \in H^2(C_3; A_3)$. We already know how to get α_1 from α, namely $\alpha_1 = \rho(\alpha)$ where $\rho : M \to M/M^{\Phi}$ is the canonical projection. Now let's work on α_2.

Let's pick a 2-cocycle $a \in \alpha$. Let $\eta \in \mathrm{Aut}_S(M)_\alpha$, so $\eta * a$ and a are cohomologous. Recall that

$$\eta * a(\sigma_1, \sigma_2) = f\left(a(A^{-1}(\sigma_1), A^{-1}(\sigma_2))\right)$$

where $\eta = (f, A)$. Let $c \in C^1(\Phi; M)$ satisfy

$$\eta * a - a = \delta c.$$

Define $\bar{\eta} : \pi \to \pi$ by

$$\bar{\eta}(\sigma, m) = (A(\sigma), f(m) + c(A(\sigma))). \tag{41}$$

Exercise 4.5: Show that $\bar{\eta} \in \mathrm{Aut}(\pi, M)$.

Lemma 4.1: *The map $X : \mathrm{Aut}_S(M)_\alpha \to B^2(\Phi; M)$ defined by $\eta \mapsto \eta * a - a = \delta c$ is a cocycle in $Z^1(\mathrm{Aut}_S(M)_\alpha; B^2(\Phi; M))$ and thus defines a class in $H^1(\mathrm{Aut}_S(M)_\alpha; B^2(\Phi; M))$ which is independent of the choice of $a \in \alpha$.*

Proof: Let's write $\mathrm{Aut}_S(M)_\alpha = C_2$ and $B^2(\Phi; M) = B^2$, so we are looking at a map X in $C^1(C_2; B^2)$, and we want to show that its coboundary is 0.

$$\begin{aligned} \delta X(\eta, \tau) &= \eta * X(\tau) - X(\eta\tau) + X(\eta) \\ &= \eta * (\tau * a - a) - \eta\tau * a + a + \eta * a - a \\ &= \eta\tau * a - \eta * a - \eta\tau * a + \eta * a \\ &= 0. \end{aligned}$$

Now the map $\eta \mapsto \eta * \delta c' - \delta c'$ is exhibited as an element of $B^1(C_2; B^2)$, so the class of X in $H^1(C_2; B^2)$ is independent of the choice of $a \in \alpha$. □

We denote the class of X in $H^1(C_2; B^2)$ by $\chi(\alpha)$.

Exercise 4.6: Show that $\chi(\alpha) = 0$ iff the exact sequence

$$0 \longrightarrow A_2 \longrightarrow B_2 \longrightarrow C_2 \longrightarrow 0$$

splits. (**Hint:** $\chi(\alpha) = 0$ implies $\eta * a = a$, so you can take $c = 0$, and then $\eta \mapsto \bar{\eta}$ is a splitting.)

This exercise suggests that we should be able to get $\alpha_2 \in H^2(C_2; A_2)$ from $\chi(\alpha) \in H^1(C_2; B^2)$, and we can. We have to get from coefficients in $B^2 = B^2(\Phi; M)$ to coefficients in $A_2 = Z^1(\Phi; M)$ in one higher dimension. This suggests we look at the exact sequence

$$0 \longrightarrow Z^1(\Phi; M) \longrightarrow C^1(\Phi; M) \xrightarrow{\delta} B^2(\Phi; M) \longrightarrow 0,$$

and take the corresponding Bockstein

$$\beta : H^1(C_2; B^2(\Phi; M)) \to H^2(C_2; Z^1(\Phi; M)).$$

Theorem 4.1 : $\beta(\chi(\alpha)) = \alpha_2$.

Proof: Let $\eta \in C_2 = \mathrm{Aut}_S(M)_\alpha$. Pick $c(\eta) \in C^1(\Phi; M)$ satisfying

$$\eta * a - a = \delta c(\eta). \tag{42}$$

We can consider c to be a map $C_2 \to C^1(\Phi; M)$, so $c \in C^1(C_2; C^1(\Phi; M))$. Now the "$\delta$" in (42) is really δ_Φ, the coboundary for the cohomology of Φ. We also have δ_{C_2}, the coboundary for the cohomology of C_2. We can consider $\delta_{C_2} c \in Z^2(C_2; C^1(\Phi; M))$. Actually we can do better since it is easy to see that

$$\delta_{C_2} c \in Z^2(C_2; Z^1(\Phi; M)).$$

The proof is exactly the same as the proof that $\delta X = 0$ in Lemma 4.1. As an element of $Z^2(C_2; C^1(\Phi; M))$, $\delta_{C_2} c$ is, of course, cohomologous to zero, but there is no reason to suppose that as an element of $Z^2(C_2; Z^1(\Phi; M))$ it is cohomologous to zero. We can take its cohomology class in the group $H^2(C_2; Z^1(\Phi; M)) = H^2(C_2; A_2)$, and, by definition, this is $\beta(\chi(\alpha))$. We can compute $\delta_{C_2} c$ easily enough. If $\eta_1, \eta_2 \in C_2 = \mathrm{Aut}_S(M)_\alpha$, then $\delta_{C_2} c(\eta_1, \eta_2) \in Z^1(\Phi; M)$, and

$$\begin{aligned} [\delta_{C_2} c(\eta_1, \eta_2)](\sigma) &= (\eta_1 * c(\eta_2))(\sigma) - c(\eta_1\eta_2)(\sigma) + c(\eta_1)(\sigma) \\ &= f_1\left[c(\eta_2)(A_1^{-1}(\sigma))\right] - c(\eta_1\eta_2)(\sigma) + c(\eta_1)(\sigma) \end{aligned} \tag{43}$$

where $\eta_1 = (f_1, A_1)$.

Now we want to compute a representative 2-cocycle for α_2. The map from C_2 to B_2 ($= \mathrm{Aut}(\pi, M)$) sending η to $\bar{\eta}$ defined by (41) is a section for $p_2 : B_2 \to C_2$ since $\bar{\eta}|M$ is f, and the map $\bar{\eta}$ induces on Φ is A. (Both of these facts are clear from (41).) If we take $\eta_1, \eta_2 \in C_2$, say $\eta_1 = (f_1, A_1)$ and $\eta_2 = (f_2, A_2)$, then a representative, a_2, for α_2 is given by

$$a_2(\eta_1, \eta_2) = \bar{\eta}_1 \bar{\eta}_2 (\overline{\eta_1\eta_2})^{-1}.$$

First we observe that

$$\bar{\eta}^{-1}(\sigma, m) = (A^{-1}(\sigma), f^{-1}(m) - f^{-1}[c(\eta)](\sigma).$$

A brief computation shows that $a_2(\eta_1, \eta_2)$ considered as an element of $\mathrm{Aut}^0(\pi)$ is given by

$$\begin{aligned} [a_2(\eta_1, \eta_2)](\sigma, m) &= \left[\bar{\eta}_1 \circ \bar{\eta}_2 \circ (\overline{\eta_1\eta_2})^{-1}\right](\sigma, m) \\ &= (\sigma, m - [c(\eta_1, \eta_2)](\sigma) + f_1[c(\eta_2)](A_1^{-1}(\sigma)) + [c(\eta_1)](\sigma)). \end{aligned} \tag{44}$$

Recall that $\mathrm{Aut}^0(\pi)$ is identified with $Z^1(\Phi;M)$ by associating a cocycle $z \in Z^1(\Phi;M)$ to the automorphism

$$(\sigma, m) \mapsto (\sigma, m + z(\sigma)).$$

This follows from Lemma 1.2. A comparison of (43) with (44) shows $\delta_{C_2} c = a_2$ as desired. □

Remarks: i) What we are really doing in the previous proposition is computing an easy differential in a spectral sequence. This should show you that spectral sequences are complicated. In the next volume (currently imaginary), we hope to restate all of this in the language of spectral sequences.

ii) Suppose M^Φ is not the center of π. Can we save some of this? The sequence (40) which is the top row of the basic diagram won't work, but we can fix it up by replacing $B_1 = \pi$ by $B_1 = \pi/M^\Phi$. Unfortunately i_A, i_B, and i_C won't, in general, be injective. However, as a non-trivial exercise you can, perhaps, prove the following:

Exercise 4.7: i) Show that Theorem 4.1 holds in this more general situation.

ii) Show that we can recover the whole basic diagram as follows: Let Z be the center of π, and replace π by π/Z. M/M^Φ is naturally isomorphic to $MZ/M^\Phi Z$, so that needs no change, but Φ must be replaced by $\Phi/Z \cap \ker j$ where $j : \Phi \to \mathrm{Aut}(M)$ is the action map.

What about α_3? Well, results here are harder to come by. Since we need some more hypotheses, it is more convenient to postpone the study of α_3 to the next section where we study automorphisms of Bieberbach groups.

5. Automorphisms of Bieberbach Groups

We now return to the situation of Section 1. We let π be a Bieberbach group, and we have the usual exact sequence

$$0 \longrightarrow M \longrightarrow \pi \xrightarrow{p} \Phi \longrightarrow 1, \tag{45}$$

and the basic 9-diagram (6) which we reproduce here for the convenience

of the reader

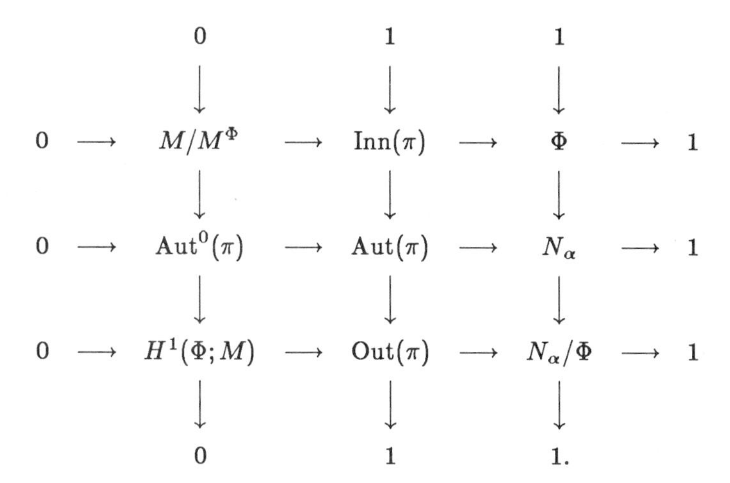

We know that the cohomology class of the top row, $\alpha_1 \in H^2(\Phi; M/M^\Phi)$, is $\rho_*(\alpha)$ where $\rho : M \to M/M^\Phi$ is the canonical projection and $\alpha \in H^2(\Phi; M)$ is the cohomology class of the extension (45). Furthermore M/M^Φ is a free abelian group of finite rank. It is also a faithful Φ-module on which Φ fixes only the identity element. We can also determine α_2, the cohomology class of the middle row, by using Theorem 4.1.

We would like to use Proposition 3.2 to determine α_3, the cohomology class of the bottom row, from α_1 and α_2, the cohomology classes of the other two rows. To use this proposition we must show that

$$\left(p_A(H^1(\Phi; \mathrm{Aut}^0(\pi)))\right)^{N_\alpha} \subset \mathrm{res}(H^1(N_\alpha; H^1(\Phi; M))), \tag{46}$$

which unfortunately we cannot, in general, show. In the special case that $H^1(\Phi; M)$ and $H^1(\Phi; M/M^\Phi)$ are isomorphic, then (46) follows from a general theorem concerning 9-diagrams. Before we give this theorem, we reproduce the general 9-diagram.

$$\begin{array}{ccccccccc}
 & & 0 & & 1 & & 1 & & \\
 & & \downarrow & & \downarrow & & \downarrow & & \\
0 & \longrightarrow & A_1 & \xrightarrow{i_1} & B_1 & \xrightarrow{p_1} & C_1 & \longrightarrow & 1 \\
 & & \downarrow i_A & & \downarrow i_B & & \downarrow i_C & & \\
0 & \longrightarrow & A_2 & \xrightarrow{i_2} & B_2 & \xrightarrow{p_2} & C_2 & \longrightarrow & 1 \\
 & & \downarrow p_A & & \downarrow p_B & & \downarrow p_C & & \\
0 & \longrightarrow & A_3 & \xrightarrow{i_3} & B_3 & \xrightarrow{p_3} & C_3 & \longrightarrow & 1 \\
 & & \downarrow & & \downarrow & & \downarrow & & \\
 & & 0 & & 1 & & 1. & &
\end{array}$$

Theorem 5.1 : *Suppose we have a short exact sequence of groups*

$$1 \longrightarrow C_1 \xrightarrow{i_C} C_2 \xrightarrow{p_C} C_3 \longrightarrow 1,$$

and a short exact sequence of C_2-modules

$$0 \longrightarrow A_1 \xrightarrow{i_A} A_2 \xrightarrow{p_A} A_3 \longrightarrow 0,$$

with A_3 a trivial C_1-module. Also suppose that A_1 and A_2 are free abelian groups of the same finite rank, that

$$A_3 = H^1(C_1; A_1), \tag{47}$$

and that C_1 is a finite group with $A_2^{C_1} = 0$. Then we have

$$\left(p_A(H^1(C_1; A_2))\right)^{C_2} \subset \operatorname{res}(H^1(C_2; A_3)).$$

Equivalently (see the discussion before Proposition 3.2), we can conclude that

$$\ker\beta \subset \ker \operatorname{con}$$

where $\beta : H^1(C_1; A_3)^{C_2} \to H^2(C_1; A_1)^{C_2}$ is the Bockstein map for the sequence

$$0 \longrightarrow A_1 \longrightarrow A_2 \longrightarrow A_3 \longrightarrow 0, \tag{48}$$

and con : $H^1(C_1; A_3)^{C_2} \to H^2(C_3; A_3)$ *is the connection for the extension*

$$1 \longrightarrow C_1 \longrightarrow C_2 \longrightarrow C_3 \longrightarrow 1.$$

Before beginning the proof, let's see which of the onerous hypotheses of the theorem are automactically verified in our situation. $A_1 = M/M^\Phi$ and $A_2 = \mathrm{Aut}^0(\pi)$ are free abelian groups of the same finite rank since $H^1(\Phi; M)$ is finite. $C_1 = \Phi$ is certainly finite and $A_2^{C_1} = (\mathrm{Aut}^0(\pi))^\Phi = Z^1(\Phi; M)^\Phi$, and we will show this is 0 after the proof of the theorem. Hence the only extra condition is the one mentioned above, namely that $H^1(\Phi; M)$ and $H^1(\Phi; M/M^\Phi)$ are isomorphic, i.e. $A_3 = H^1(C_1; A_1)$.

Proof: The exact sequence (48) gives rise to long exact cohomology sequences for C_1 and C_2. By considering the restriction from C_2 to C_1 we get the following "ladder":

$$\begin{array}{ccc}
0 & & 0 \\
\downarrow & & \downarrow \\
H^0(C_2; A_3) & \xrightarrow{\text{res}} & H^0(C_1; A_3) \\
\downarrow \beta_2 & & \downarrow \beta_1 \\
H^1(C_2; A_1) & \xrightarrow{\text{res}} & H^1(C_1; A_1) \\
\downarrow (i_A)_2 & & \downarrow (i_A)_1 \\
H^1(C_2; A_2) & \xrightarrow{\text{res}} & H^1(C_1; A_2) \\
\downarrow (p_A)_2 & & \downarrow (p_A)_1 \\
H^1(C_2; A_3) & \xrightarrow{\text{res}} & H^1(C_1; A_3) \\
\downarrow \beta_2 & & \downarrow \beta_1 \\
\vdots & & \vdots
\end{array}$$

where we have appended subscripts "1" and "2" to attempt to avoid confusion. The sequences begin with 0's since $H^0(C_1; A_2) = A_2^{C_1} = 0$ by hypothesis, and $H^0(C_2; A_2) = A_2^{C_2} = 0$ because if C_1 does not fix any non-trivial points of A_2, C_2 certainly doesn't.

We can use the extension of the Hochschild–Serre exact sequence (Theorem 2.1) given in Equation (25) to see that the kernels of the second and third horizontal arrows (i.e. the restrictions) are $H^1(C_3; A_1^{C_1})$ and $H^1(C_3; A_2^{C_1})$ respectively. The condition $A_2^{C_1} = 0$ implies that both of these cohomology groups are 0. Also $H^1(C_1; A_1) = A_3$ is a finite group, so β_1 is a monomorphism from A_3 to itself, and hence must be a isomorphism which implies that $(i_A)_1 = 0$. Consequently β_2 is an isomorphism and $(i_A)_2 = 0$. Thus all the horizontal arrows (i.e. the restrictions) are monomorphisms, and the first two have image equal to $A_3^{C_2}$. In addition, $(p_A)_1$ and $(p_A)_2$ are monomorphisms.

Now we continue the ladder down and, since $(p_A)_1$ and $(p_A)_2$ are monomorphisms, we can start with 0's on the left.

$$\begin{array}{ccc}
0 & & 0 \\
\downarrow & & \downarrow \\
H^1(C_2; A_2) & \xrightarrow{\text{res}} & H^1(C_1; A_2) \\
\downarrow{\scriptstyle (p_A)_2} & & \downarrow{\scriptstyle (p_A)_1} \\
H^1(C_2; A_3) & \xrightarrow{\text{res}} & H^1(C_1; A_3) \\
\downarrow{\scriptstyle \beta_2} & & \downarrow{\scriptstyle \beta_1} \\
H^2(C_2; A_1) & \xrightarrow{\text{res}} & H^2(C_1; A_1) \\
\downarrow{\scriptstyle (i_A)_2} & & \downarrow{\scriptstyle (i_A)_1} \\
H^2(C_2; A_2) & \xrightarrow{\text{res}} & H^2(C_1; A_2) \\
\vdots & & \vdots
\end{array}$$

Using the Hochschild–Serre exact sequence (25) and reasoning as above we can conclude that the first horizontal arrow is a monomorphism with image $H^1(C_1; A_2)^{C_2}$ and that the second horizontal arrow has kernel isomorphic to $H^1(C_3; A_3)$. In any event, Lemma 2.1 tells us that the images of all the horizontal maps lie in the fixed points of C_2. We would therefore like to change the right column by taking fixed points under the action of C_2 but we won't get exactness everywhere.

Exercise 5.1: Show that after taking fixed points under the action of C_2 the cohomology sequence is still exact at $H^1(C_1; A_3)^{C_2}$.

We therefore obtain the following exact and commutative diagram:

$$\begin{array}{ccccccc}
 & & & & 0 & & \\
 & & & & \downarrow & & \\
 & & 0 & & H^1(C_3;A_3) & & \\
 & & \downarrow & & \downarrow \text{inf} & & \\
0 & \longrightarrow & H^1(C_2;A_2) & \overset{(p_A)_2}{\longrightarrow} & H^1(C_2;A_3) & \overset{\beta_2}{\longrightarrow} & H^2(C_2;A_1) \\
 & & \downarrow \text{res} & & \downarrow \text{res} & & \downarrow \text{res} \\
0 & \longrightarrow & H^1(C_1;A_2)^{C_2} & \overset{(p_A)_1}{\longrightarrow} & H^1(C_1;A_3)^{C_2} & \overset{\beta_1}{\longrightarrow} & H^2(C_1;A_1)^{C_2} \\
 & & \downarrow & & \downarrow \text{con} & & \\
 & & 0 & & H^2(C_3;A_3) & & \\
 & & & & \downarrow \text{inf} & & \\
 & & & & H^2(C_2;A_3). & &
\end{array}$$

Exercise 5.2: Check that all the exactness, commutativity, and 0's in the above diagram are justified.

It is now an easy diagram chase to see that

$$\ker\beta_1 \subset \ker \text{con},$$

as desired. □

Now we give the promised result that $A_2^{C_1} = 0$ in our situation. Recall that $A_2^{C_1} = Z^1(\Phi; M)^\Phi$.

Lemma 5.1: *Given the extension (45), we have* $Z^1(\Phi; M)^\Phi = 0$.

Proof: Let $Z \in Z^1(\Phi; M)$ and $\lambda \in \Phi$. It is easy to see that $\lambda * z(\sigma) = z(\sigma\lambda) - z(\lambda)$. (Look at $\delta z(\lambda, \lambda^{-1}\sigma\lambda)$.) If $\lambda * z = z$, $2z(\sigma) = z(\sigma^2)$, so by induction $2^n z(\sigma) = z(\sigma^{2^n})$. Since Φ is finite and M is free, we can conclude that $z = 0$. □

For all of the above to work, we still must have $H^1(\Phi; M)$ be isomorphic to $H^1(\Phi; M/M^\Phi)$. We can only say a little about when this is the case.

Exercise 5.3: i) Show that $H^1(\Phi; M)$ is isomorphic to a subgroup of $H^1(\Phi; M/M^\Phi)$. (**Hint:** Look at the long exact cohomology sequence for the chohmology of Φ associated to the exact sequence

$$0 \longrightarrow M^\Phi \longrightarrow M \longrightarrow M/M^\Phi \longrightarrow 0.$$

ii) Show that if Φ is a perfect group or if M^Φ is a Φ-direct summand of M, then $H^1(\Phi; M)$ is isomorphic to $H^1(\Phi; M/M^\Phi)$.

Here's another exercise which is an easy corollary to the above theorem.

Exercise 5.4: i) If $H^1(\Phi; M)$ and $H^1(\Phi; M/M^\Phi)$ are isomorphic, then $\alpha_1 = 0$ implies $\alpha_2 = \alpha_3 = 0$.

ii) Show that $\alpha_1 = 0$ is Φ if cyclic.

We finish this section with some examples.

Example 5.1: i) Let Φ have order two, and take M to be the group ring $\mathbb{Z}[\Phi]$ with Φ acting, as usual, by left multiplication. Then $H^1(\Phi; M) = 0$ while it is easily seen that $H^1(\Phi; M/M^\Phi)$ has order two.

ii) Let Φ be the symmetric group on three symbols (i.e. the dihedral group of order six), and take for M the free abelian group of rank four with generators a, b, c, and d. Let ρ and τ be generators of Φ of orders three and two respectively. Let ρ and τ act on M by the matrices

$$\begin{pmatrix} 0 & -1 & 0 & 0 \\ 1 & -1 & 0 & 0 \\ 0 & 0 & 1 & 0 \\ 0 & 0 & 0 & 1 \end{pmatrix} \text{ and } \begin{pmatrix} 0 & 1 & 0 & 0 \\ 1 & 0 & 0 & 0 \\ 0 & 0 & 1 & 1 \\ 0 & 0 & 0 & -1 \end{pmatrix}.$$

Consider the group π formed by the usual multiplication on $\Phi \times M$ where we do not yet specify the cocycle $f \in Z^2(\Phi; M)$.

Exercise 5.5: Show that the cocycle f is completely determined by the conditions $(\rho, 0)^3 = (1, d)$ and $(\tau, 0)^2 = (1, c)$ (and, of course, the relations in Φ and the above action of Φ on M).

This group π is the fundamental group of a flat 4-dimensional manifold whose holonomy group is Φ. Since M is the direct sum of three Φ-modules, $H^1(\Phi; M)$ breaks up into the direct sum of three groups.

Exercise 5.6: i) Show that $H^1(\Phi; M)$ has order two and this is the contribution of the third direct summand of M, the one generated by d.

ii) Let R be the cyclic subgroup of order three generated by ρ in Φ. Using the notation of this section, show that i_A induces an isomorphism of $H^2(R; M/M^\Phi)$ with $H^2(R; \mathrm{Aut}^0(\pi))$.

iii) Then show that $i_A(\mathrm{res}(\alpha_1)) \neq 0$.

iv) Finally use Theorem 4.1 to conclude that $\alpha_2 \neq 0$.

6. Automorphisms of Flat Manifolds

Recall that if X and Y are riemannian manifolds, then a diffeomorphism $f : X \to Y$ is called an *affinity* or *affine equivalence* if it preserves the riemannian connection. We denote the group of affinities from X to itself by *Aff(X)*. It is known to be a Lie group (see [49], page 229) although that fact will not be of much importance here. The aim of this section is to give some idea of what $\mathrm{Aff}(X)$ is in the case X is a (compact) flat (riemannian) manifold. This material and that of the next two sections is taken from [20] and [21].

As usual, we let $\pi = \pi_1(X)$ and regard it as a subgroup of $\mathcal{M}_n$.

Lemma 6.1: *Let $N(\pi)$ be the normalizer of π in $\mathcal{A}_n$. Then we have an exact sequence*

$$1 \longrightarrow \pi \longrightarrow N(\pi) \xrightarrow{f} \mathrm{Aff}(X) \longrightarrow 1.$$

Proof: We define f in the obvious fashion, i.e. given $x \in X$, lift it to $\mathbb{R}^n$, map it via an element of $N(\pi)$, and then project back to X. First we show that f is surjective. Let $\psi \in \mathrm{Aff}(X)$ and let $p : \mathbb{R}^n \to X$ be the projection. Since $\mathbb{R}^n$ covers X, there is $\tilde{\psi} : \mathbb{R}^n \to \mathbb{R}^n$ s.t. $\tilde{\psi}$ covers ψ, i.e.

$$\begin{array}{ccc} \mathbb{R}^n & \xrightarrow{\tilde{\psi}} & \mathbb{R}^n \\ \downarrow p & & \downarrow p \\ X & \xrightarrow{\psi} & X \end{array}$$

is commutative. Clearly $f(\tilde{\psi}) = \psi$. It remains to show that $\tilde{\psi} \in N(\pi)$.

Since p is a local affinity, $\tilde{\psi} \in \mathcal{A}_n$. Now take $\sigma \in \pi$ and $x_0 \in \mathbb{R}^n$. Then

$$p[\tilde{\psi}(\sigma \cdot x_0)] = \psi[p(\sigma \cdot x_0)] = \psi[p(x_0)] = p[\tilde{\psi}(x_0)],$$

so $\tilde{\psi}(\sigma \cdot x_0)$ is in the same orbit as $\tilde{\psi}(x_0)$. Hence $\exists \sigma' \in \pi$ s.t. $\tilde{\psi}(\sigma \cdot x_0) = \sigma' \cdot \tilde{\psi}(x_0)$. Then $\tilde{\psi}\sigma$ and $\sigma'\tilde{\psi}$ both cover ψ and agree at x_0, so we must have $\tilde{\psi}\sigma = \sigma'\tilde{\psi}$ or $\tilde{\psi}\tilde{\psi}^{-1} = \sigma' \in \pi$. Therefore $\tilde{\psi} \in N(\pi)$.

To see that $\ker f = \pi$ suppose that $\tilde{\psi} \in N(\pi)$ induces the identity on X. Covering space theory (see [55], Chapter V) tells us that $\tilde{\psi} \in \pi$, while clearly any element of π induces the identity on X. □

Lemma 6.2: *Let $C(\pi)$ be the centralizer of π in $\mathcal{A}_n$. Then there is an exact sequence*

$$1 \longrightarrow C(\pi) \longrightarrow N(\pi) \longrightarrow \mathrm{Aut}(\pi) \longrightarrow 1.$$

Proof: The only thing that is not obvious is that the map from $N(\pi)$ to $\mathrm{Aut}(\pi)$ is an epimorphism. This map is, of course, simply the one induced by conjugation, and the second Bieberbach Theorem (Theorem 4.1 of Chapter I) says precisely that an abstract isomorphism between Bieberbach groups can be realized by conjugation within $\mathcal{A}_n$. □

Lemma 6.3: *Let Z be the center of π, and $M = \pi \cap \mathbb{R}^n$. Then*

i) $C(\pi) \subset \mathbb{R}^n$,

ii) $C(\pi) = M^\Phi \otimes \mathbb{R} = (\mathbb{R}^n)^\Phi$, *and*

iii) $Z = C(\pi) \cap \pi = (M^\Phi \otimes \mathbb{R}) \cap \pi = M^\Phi$.

Proof: Let $(m, s) \in C(\pi)$ and $(I, t) \in M$. Then

$$(m, s)(I, t) = (m, mt + s),$$

and

$$(I, t)(m, s) = (m, s + t).$$

Hence $mt = t$. Since M spans $\mathbb{R}^n$ by the First Bieberbach Theorem (Theorem 3.1 of Chapter I), m must be I, and i) is proved.

Now let $(I, s) \in C(\pi)$ and $(m, t) \in \pi$. Then

$$(m, t)(I, s) = (m, ms + t),$$

and

$$(I, s)(m, t) = (m, s + t).$$

Hence $ms = m \; \forall m \in \Phi$. Therefore $s \in (\mathbb{R}^n)^\Phi$. So for ii), it only remains to show that $M^\Phi \otimes \mathbb{R} = (\mathbb{R}^n)^\Phi$.

Clearly $M^\Phi \otimes \mathbb{R} \subset (\mathbb{R}^n)^\Phi$. Suppose $M^\Phi \otimes \mathbb{R} \neq (M^\Phi \otimes \mathbb{R})^\Phi (= (\mathbb{R}^n)^\Phi)$. Then $(M/M^\Phi \otimes \mathbb{R})^\Phi \neq 0$, i.e. if one considers Φ acting on M/M^Φ, then the space $\bigcap_{\sigma\in\Phi} \ker(\sigma - I) \otimes 1 \neq 0$. Now the matrix of σ acting on M/M^Φ is a rational matrix $\forall \sigma \in \Phi$, so there is a rational vector in $(M/M^\Phi \otimes \mathbb{R})^\Phi$, and hence an integral vector, i.e. $\exists \tilde{a} \in M/M^\Phi$ s.t. $\sigma \cdot \tilde{a} = \tilde{a}\ \forall \sigma \in \Phi$.

If we consider Φ as acting on M, $\exists a \in M$ s.t. $\sigma \cdot a - a \in M^\Phi\ \forall \sigma \in \Phi$. Let $b = \sigma \cdot a - a \in M^\Phi$. We get

$$\sigma^2 \cdot a - \sigma \cdot a = \sigma \cdot b = b,$$

so

$$\sigma^2 \cdot a = \sigma \cdot a + b = a + 2b, \ldots$$

and

$$\sigma^k \cdot a = a + kb.$$

Since Φ is finite, $\sigma^k = I$ for some k, so $kb = 0$ or $b = 0$ and $a \in M^\Phi$ (compare the proof of Lemma 5.1). Finally we get $a = 0$, and $(M/M^\Phi \otimes \mathbb{R})^\Phi = 0$, so $(M^\Phi \otimes \mathbb{R}) = (\mathbb{R}^n)^\Phi$ which proves ii).

iii) is trivial by ii). $\square$

Recall that if $\mathrm{Inn}(\pi)$ is the group of inner automorphisms of π, then the group of outer "automorphisms" is defined by $\mathrm{Out}(\pi) = \mathrm{Aut}(\pi)/\mathrm{Inn}(\pi)$.

Lemma 6.4: *$\mathrm{Aff}(X)/f((\mathbb{R}^n)^\Phi)$ is isomorphic to $\mathrm{Out}(\pi)$.*

Proof: We have the following diagram of exact sequences:

$$\begin{array}{ccccccccc}
 & & & & 1 & & & & \\
 & & & & \downarrow & & & & \\
 & & & & C(\pi) & = & (\mathbb{R}^n)^\Phi & & \\
 & & & & \downarrow & & & & \\
1 & \longrightarrow & \pi & \longrightarrow & N & \xrightarrow{f} & \mathrm{Aff}(X) & \longrightarrow & 1 \\
 & & & & \downarrow{\scriptstyle g} & & & & \\
 & & & & \mathrm{Aut}(\pi) & & & & \\
 & & & & \downarrow & & & & \\
 & & & & 1. & & & &
\end{array}$$

Since $g(\pi)$ is clearly $\mathrm{Inn}(\pi)$, the lemma follows by diagram chasing. $\square$

Let $\mathrm{Iso}(X)$ be the group of isometries of X. It is a famous theorem of riemannian geometry (due to Meyers and Steenrod) that $\mathrm{Iso}(X)$ is a Lie group (see [49], page 239). Recall that if G is a topological group, G_0 is its identity component.

Theorem 6.1 : $\mathrm{Aff}_0(X)$ *is isomorphic to* $f((\mathbb{R}^n)^\Phi)$. *Thus we can conclude that* $\mathrm{Aff}(X)/\mathrm{Aff}_0(X)$ *is isomorphic to* $\mathrm{Out}(\pi)$.

Proof: It is a general theorem that $\mathrm{Iso}_0(X) = \mathrm{Aff}_0(X)$ if X is any compact riemannian manifold (see [49], page 244). Thus it suffices to show that $\mathrm{Iso}_0(X)$ is isomorphic to $f((\mathbb{R}^n)^\Phi)$.

Now all of the elements of $(\mathbb{R}^n)^\Phi$ are translations (in $\mathcal{A}_n$), hence they are isometries, and $f((\mathbb{R}^n)^\Phi) \subset \mathrm{Iso}(X)$. Since $(\mathbb{R}^n)^\Phi$ is connected, $f((\mathbb{R}^n)^\Phi) \subset \mathrm{Iso}_0(X)$. It remains to show that $f((\mathbb{R}^n)^\Phi)$ is all of $\mathrm{Iso}_0(X)$. By the general theory of Lie groups , it suffices to show that each 1-parameter subgroup $\{\phi_t\}$ of $\mathrm{Iso}_0(X)$ is the f image of a 1-parameter subgroup of $(\mathbb{R}^n)^\Phi$.

Let V be the (Killing) vector field on X corresponding to $\{\phi_t\}$. Let $\tilde{V}$ be the corresponding vector field on $\mathbb{R}^n$ which is the universal cover of X.

Exercise 6.1: Show that $\tilde{V}$ corresponds to a 1-parameter group, $\{\tilde{\phi}_t\}$ of diffeomorphisms of $\mathbb{R}^n$.

Thus $\tilde{\phi}_t \in N(\pi)$. Since π is discrete, the map from π to π which sends α to $\tilde{\phi}_t \alpha \tilde{\phi}_t^{-1}$ is independent of t. Since $\tilde{\phi}_0$ is the identity, it must follow that $\tilde{\phi}_t$ is in $C(\pi) = \mathbb{R}^n$ which is contained in the group of translations. Thus $\{\phi_t\}$ is a 1-parameter group of translations, and the theorem follows. □

Remark: We have used a number of results from riemannian geometry in the above arguments without proof. When you look at the references mentioned above, you should also look for results that will help you prove

Exercise 6.2: Show that $\mathrm{Aff}_0(X)$ is a torus whose dimension equals the first Betti number of X, i.e.

$$\mathrm{Aff}_0(X) \approx (S^1)^{\beta_1}$$

where β_1 is the dimension of $H^1(X; \mathbb{R}^n)$.

The upshot of this is that $\mathrm{Aff}(X)_0$ is not very interesting, so we can concentrate on $\mathrm{Aff}(X)/\mathrm{Aff}_0(X)$ which we have seen is isomorphic to $\mathrm{Out}(\pi)$.

We end this section with two interesting examples.

Example 6.1: Recall the Hantsche–Wendt group described in Example 2.4 of Chapter I. We let X be the corresponding three-dimensional flat manifold. Naturally X is called the *Hantsche–Wendt manifold.* We are going to compute the group of affinities of X.

It is easy to check that X is orientable and has holonomy group $\Phi(X)$ isomorphic to $\mathbb{Z}_2 \times \mathbb{Z}_2$ (remember we think of Φ multiplicatively), which is the Klein 4-group. Exercise 2.2 of Chapter I tells us that $H_1(X;\mathbb{Z})(\approx \pi/[\pi,\pi]) \approx \mathbb{Z}_4 \oplus \mathbb{Z}_4$. Thus the first Betti number of X is 0, so $\text{Aff}(X) \approx \text{Out}(\pi_1(X))$. Let $\pi = \pi_1(X)$ and $\Phi = \Phi(X)$.

We want to describe π rather carefully. Let $M = \mathbb{Z} \oplus \mathbb{Z} \oplus \mathbb{Z}$. Thus $\text{Aut}(M) = J_3$, the 3-dimensional unimodular group. We denote the three elements of order two of Φ by σ_1, σ_2, and σ_3. Define $j : \Phi \to \text{Aut}(M)$ by $j(1) = I$ and

$$j(\sigma_1) = \begin{pmatrix} +1 & 0 & 0 \\ 0 & -1 & 0 \\ 0 & 0 & -1 \end{pmatrix},$$

$$j(\sigma_2) = \begin{pmatrix} -1 & 0 & 0 \\ 0 & +1 & 0 \\ 0 & 0 & -1 \end{pmatrix},$$

and

$$j(\sigma_3) = \begin{pmatrix} -1 & 0 & 0 \\ 0 & -1 & 0 \\ 0 & 0 & +1 \end{pmatrix}.$$

Let N be the normalizer of $j(\Phi)$ in $\text{Aut}(M)$. Let S_3 be the subgroup of $\text{Aut}(M)$ consisting of permutation matrices. Clearly $S_3 \subset N$.

Exercise 6.3: Consider the map $\rho : N \to \text{Aut}(\Phi)$ given by conjugation. Show that S_3 is mapped isomorphically onto $\text{Aut}(\Phi)$ by ρ.

We view M as a Φ-module by means of j. Let Φ_i be the subgroup of Φ generated by σ_1 for $i = 1,2,3$. Put $M_i = M^{\Phi_i}$. A glance at the matrices above shows that $M_1 = \mathbb{Z} \oplus 0 \oplus 0, M_2 = 0 \oplus \mathbb{Z} \oplus 0$, and $M_3 = 0 \oplus 0 \oplus \mathbb{Z}$.

Clearly $\ker\rho = C(j(\Phi))$, the centralizer of Φ in $\text{Aut}(M)$. Thus if

$A \in \ker\rho, A(M_i) \subset M_i$, i.e. A is a diagonal matrix. Hence

$$\ker\rho = \begin{pmatrix} \pm 1 & 0 & 0 \\ 0 & \pm 1 & 0 \\ 0 & 0 & \pm 1 \end{pmatrix} \approx \mathbb{Z}_2 \times \mathbb{Z}_2 \times \mathbb{Z}_2.$$

We can summarize this as follows: The exact sequence

$$0 \longrightarrow \ker\rho \longrightarrow N \stackrel{\rho}{\longrightarrow} \mathrm{Aut}(\Phi) \longrightarrow 1$$

splits, and thus N is isomorphic to the semi-direct product of $S_3 (\approx \mathrm{Aut}(\Phi))$ acting on $\mathbb{Z}_2 \times \mathbb{Z}_2 \times \mathbb{Z}_2 (\approx \ker\rho)$ by permuting the coordinates.

We now examine $H^2(\Phi; M)$. Notice that the M_i are actually Φ-submodules, and $M = M_1 \oplus M_2 \oplus M_3$. Thus

$$H^2(\Phi; M) = H^2(\Phi; M_1) \oplus H^2(\Phi; M_2) \oplus H^2(\Phi; M_3).$$

It follows from Proposition 5.1 of Chapter I (and several other places) that $H^2(\Phi; M_i) \approx \mathbb{Z}_2$ so,

$$H^2(\Phi; M) \approx \mathbb{Z}_2 \oplus \mathbb{Z}_2 \oplus \mathbb{Z}_2.$$

Since Φ_i acts trivially on M_i, $H^2(\Phi_i; M_i) \approx \mathbb{Z}_2$ also. In fact, the natural map $H^2(\Phi; M) \to H^2(\Phi_i; M)$ is an isomorphism.

Let $\alpha \in H^2(\Phi; M)$ be the element all of whose coordinates are non-trivial. Let π be the corresponding group extension,

$$0 \longrightarrow M \longrightarrow \pi \longrightarrow \Phi \longrightarrow 1.$$

Exercise 6.4: Show that this π is the Hantsche–Wendt group of Example 2.4 of Chapter I.

We now ask how N acts on $H^2(\Phi; M)$. For this it is perhaps worthwhile making the isomorphism $\theta : H^2(\Phi; M) \approx \mathbb{Z}_2 \oplus \mathbb{Z}_2 \oplus \mathbb{Z}_2$ more explicit. Let $g : \Phi \times \Phi \to M$ be a 2-cocycle and $[g] \in H^2(\Phi; M)$ the corresponding cohomology class. Then the ith coordinate of $\theta([g])$ is the ith coordinate modulo 2 of $g(\sigma_i, \sigma_i) \in M = M_1 \oplus M_2 \oplus M_3$. It should now be clear that $\ker\rho \subset N$ acts trivially on $H^2(\Phi; M)$, and that $N/\ker\rho \approx S_3$ acts by permuting the coordinates. Thus we can see that α is fixed by all of N so $N_\alpha = N$, and $N_\alpha/\Phi = N/\Phi$. From the exact sequence

$$1 \longrightarrow \ker\rho/\Phi \longrightarrow N/\Phi \longrightarrow n/\ker\rho \longrightarrow 1$$

and the splitting of the exact sequence

$$1 \longrightarrow \ker\rho \longrightarrow N \longrightarrow N/\ker\rho \longrightarrow 1,$$

we can conclude that N_α/Φ is isomorphic to $\mathbb{Z}_2 \times S_3$.

Now we compute $H^1(\Phi; M)$. As before

$$H^1(\Phi; M) = H^1(\Phi; M_1) \oplus H^1(\Phi; M_2) \oplus H^1(\Phi; M_3).$$

Since Φ_i acts trivially on M_i, Φ/Φ_i acts on M_i, and an easy calculation shows that

$$H^1(\Phi; M_i) \approx H^1(\Phi/\Phi_i; M_i) \approx \mathbb{Z}_2.$$

As above , N acts on $H^1(\Phi; M) \approx \mathbb{Z}_2 \oplus \mathbb{Z}_2 \oplus \mathbb{Z}_2$, $\ker\rho$ acts trivially, and $N/\ker\rho \approx S_3$ acts via permutations.

What we have done at this point is to compute the last row of the basic 9-diagram for this group. We summarize the results in the following diagram:

$$\begin{array}{ccccccccc} 0 & \longrightarrow & H^1(\Phi; M) & \longrightarrow & \text{Out}(\pi) & \longrightarrow & N_\alpha/\Phi & \longrightarrow & 1 \\ & & \downarrow\approx & & \downarrow = & & \downarrow\approx & & \\ 0 & \longrightarrow & \mathbb{Z}_2 \oplus \mathbb{Z}_2 \oplus \mathbb{Z}_2 & \longrightarrow & \text{Out}(\pi) & \longrightarrow & \mathbb{Z}_2 \times S_3 & \longrightarrow & 1. \end{array}$$

Furthermore the action of $\mathbb{Z}_2 \times S_3$ on $\mathbb{Z}_2 \oplus \mathbb{Z}_2 \oplus \mathbb{Z}_2$ is described implicitly above. To wit, $\mathbb{Z}_2 \times 1$ acts trivially while $1 \times S_3$ acts by permutation of coordinates.

It remains to describe the class $\alpha_3 \in H^2(N_\alpha; H^1(\Phi; M))$ which corresponds to the bottom row. We can expect this to be a mess. Firstly we can say that $M^\Phi = \{0\}$, so $H^1(\Phi; M) = H^1(\Phi; M/M^\Phi)$, and the theory of Section 5 applies. We have seen, however, that this is very complex. To see what the group $\text{Aff}(X)$ really is it is easier to look at what the elements of $\text{Aff}(X)$ really do.

Exercise 6.5: Let ξ, η, and ς be the generators of the subgroup $\mathbb{Z}_2 \oplus \mathbb{Z}_2 \oplus \mathbb{Z}_2$ of $\text{Aff}(X)$. Show that these act on a point $(x, y, z) \in \mathbb{R}^3$ as follows:

$$\xi \cdot (x, y, z) = (x + 1/2, y, z),$$
$$\eta \cdot (x, y, z) = (x, y + 1/2, z)$$

and

$$\varsigma \cdot (x, y, z) = (x, y, z + 1/2).$$

Further show that the elements of the S_3 subgroup of $\mathrm{Aff}(X)$ act by permuting the coordinates, and finally show that the generator of the last $\mathbb{Z}_2$ subgroup of $\mathrm{Aff}(X)$ can be taken to act as

$$\begin{pmatrix} -1 & 0 & 0 \\ 0 & +1 & 0 \\ 0 & 0 & +1 \end{pmatrix}, \text{ or } \begin{pmatrix} +1 & 0 & 0 \\ 0 & -1 & 0 \\ 0 & 0 & +1 \end{pmatrix}, \text{ or } \begin{pmatrix} +1 & 0 & 0 \\ 0 & +1 & 0 \\ 0 & 0 & -1 \end{pmatrix},$$

$$\text{or } \begin{pmatrix} -1 & 0 & 0 \\ 0 & -1 & 0 \\ 0 & 0 & -1 \end{pmatrix}.$$

These matrices all induce the same map on X.

It is now also easy from the above exercise to determine the subgroup of $\mathrm{Aff}(X)$ which preserve orientation. It is the semi-direct product of S_3 and $\mathbb{Z}_2 \oplus \mathbb{Z}_2 \oplus \mathbb{Z}_2$ with the permutation action, i.e. the *wreath product* of S_3 and $\mathbb{Z}_2$. $\mathrm{Aff}(X)$ is this group with an additional factor of $\mathbb{Z}_2$. So X has 96 affine equivalences and 48 of these preserve orientation.

Example 6.2: If the first Betti number of X is 0, then $\mathrm{Aff}(X)$ is discrete. In the example above, we have seen that it is also finite. It has been conjectured that this is always the case, i.e. if $\beta_1 = 0$, then $\mathrm{Aff}(X)$ is finite. This example shows that this is not true.

X will be a non-orientable 4-dimensional manifold with $\Phi(X) = \mathbb{Z}_2 \times \mathbb{Z}_2$. Let σ_1, σ_2, and σ_3 be the non-trivial elements of $\Phi = \Phi(X)$. Put $M = \mathbb{Z}^4 = \mathbb{Z} \oplus \mathbb{Z} \oplus \mathbb{Z} \oplus \mathbb{Z}$. now define $j : \Phi \to \mathrm{Aut}(M)$ by $j(1) = I$ and

$$j(\sigma_1) = \begin{pmatrix} -1 & 0 & 0 & 0 \\ 0 & -1 & 0 & 0 \\ 0 & 0 & -1 & 0 \\ 0 & 0 & 0 & +1 \end{pmatrix},$$

$$j(\sigma_2) = \begin{pmatrix} -1 & 0 & 0 & 0 \\ 0 & -1 & 0 & 0 \\ 0 & 0 & +1 & 0 \\ 0 & 0 & 0 & -1 \end{pmatrix},$$

and

$$j(\sigma_3) = \begin{pmatrix} +1 & 0 & 0 & 0 \\ 0 & +1 & 0 & 0 \\ 0 & 0 & -1 & 0 \\ 0 & 0 & 0 & -1 \end{pmatrix}.$$

Exercise 6.6: Show that

$$H^2(\Phi; M) \approx \mathbb{Z}_2 \oplus \mathbb{Z}_2 \oplus \mathbb{Z}_2 \oplus \mathbb{Z}_2.$$

Let $\alpha \in H^2(\Phi; M)$ be the element whose second coordinate is 0 while all of its other coordinates are 1, and let π be the corresponding group extension.

Exercise 6.7: i) Show that π is a Bieberbach group.

ii) Show that $H_1(\pi; \mathbb{Z}) \approx \mathbb{Z}_2 \oplus \mathbb{Z}_4 \oplus \mathbb{Z}_4$.

iii) Let N be the normalizer of $j(\Phi)$ in $\mathrm{Aut}(M)$. Show that $\psi \in N$ iff

$$\psi = \begin{pmatrix} a & b & 0 & 0 \\ e & f & 0 & 0 \\ 0 & 0 & k & 0 \\ 0 & 0 & 0 & p \end{pmatrix}$$

with $a, b, e, f, k, p \in \mathbb{Z}$ and

$$\det \begin{pmatrix} a & b \\ e & f \end{pmatrix} = \pm 1,$$

and $k = p = \pm 1$.

Hence if X is the flat manifold with $\pi(X) \approx \pi$, then X has first Betti number equal zero, and we know that $\mathrm{Aff}(X) \approx \mathrm{Out}(\pi)$.

As before we can see that N acts on $H^2(\Phi; M)$ in the obvious way.

Exercise 6.8: Show that $\psi \in N_\alpha$ iff $a \equiv 1 \pmod 2$, $e \equiv 0 \pmod 2$, $k \equiv 1 \pmod 2$, and $p \equiv 1 \pmod 2$.

Let

$$\psi_n = \begin{pmatrix} 2n+1 & 1 & 0 & 0 \\ 2n & 1 & 0 & 0 \\ 0 & 0 & 1 & 0 \\ 0 & 0 & 0 & 1 \end{pmatrix}.$$

Clearly $\psi_n \in N_\alpha\ \forall n$, so N_α is infinite. Since Φ is finite, N_α/Φ is infinite, and since $\mathrm{Aff}(X)$ maps onto N_α/Φ, $\mathrm{Aff}(X)$ is infinite as claimed.

7. Automorphisms of $\mathbb{Z}_p$-manifolds

As the title indicates, this section is essentially another example. We will show that by using the results of the previous sections, it is possible to say a great deal about the automorphisms of a $\mathbb{Z}_p$-manifold. We restrict to the case of dimension p since that is the most interesting case. Recall the following classification theorem which follows from the results Chapter IV (especially Section 7). As before, C_p denotes the ideal class group of the cyclotomic ring, $\mathbb{Z}[\varsigma]$, of pth roots of unity.

Theorem 7.1 : *To every p-dimensional $\mathbb{Z}_p$-manifold X there corresponds an ideal class $\mathfrak{A}(X) \in C_p$ and X is affinely equivalent to X' iff $\mathfrak{A}(X) = \mathfrak{A}(X')$ or $\bar{\mathfrak{A}}(X) = \mathfrak{A}(X')$. Furthermore each ideal class $\mathfrak{A} \in C_p$ corresponds to some p-dimensional $\mathbb{Z}_p$-manifold.*

It also follows from Exercise 7.1 of Chapter IV that $H^1(X;\mathbb{Z}) \approx \mathbb{Z}\oplus\mathbb{Z}_p$. Hence the first Betti number of X is one, and $\mathrm{Aff}(X)$ is a circle, so we can concentrate on $\mathrm{Aff}(X)/\mathrm{Aff}_0(X)$ which we will denote by $A(X)$. We also assume that $p \neq 2$. In the case $p = 2$, X would have to be the Klein Bottle, and the results on automorphisms can be worked out as an exercise. If we allowed $p = 2$ the notation would be somewhat complicated.

To get started, we assume that $\mathfrak{A}(X)$ is the identity element of C_p, i.e. $\mathfrak{A}(X)$ is the principal ideal class. Let π be the fundamental group of X. Let $R = \mathbb{Z}[\varsigma]$. Let $\mathbb{Z}$ act on R by $n \cdot r = \varsigma^n r$ for $n \in \mathbb{Z}$ and $r \in R$.

Exercise 7.1: Show that π is isomorphic to the split extension $\mathbb{Z} \cdot R$ formed using this action.

As usual π also satisfies the exact sequence

$$0 \longrightarrow M \longrightarrow \pi \longrightarrow \Phi \longrightarrow 1.$$

Exercise 7.2: Show that M can be taken to be the subgroup $p\mathbb{Z} \cdot R \subset \pi$.

We identify Φ with the pth roots of unity in R.

Exercise 7.3: i) Show that the action of Φ on M given by the above exact sequence can be described by saying that Φ acts via multiplication on R and trivially on $p\mathbb{Z}$.

ii) Show that as submodules of M,

$$R = \{m \in M : (\sum_{\sigma \in \Phi} \cdot m = 0\}$$

and

$$p\mathbb{Z} = \{m \in M : \sigma m = m \ \forall \sigma \in \Phi\}.$$

iii) Let $j : \Phi \to \mathrm{Aut}(M)$ be the action map, and $\psi \in N$, the normalizer of $j(\Phi)$ in $\mathrm{Aut}(M)$. Show that $\psi(R) \subset R$ and $\psi(p\mathbb{Z}) \subset p\mathbb{Z}$.

Lemma 7.1: *There is a split exact sequence*

$$1 \longrightarrow Z \longrightarrow N \underset{\longleftarrow}{\overset{\longrightarrow}{}} \mathrm{Aut}(\Phi) \longrightarrow 1$$

where Z is the centralizer of $j(\Phi)$ in $\mathrm{Aut}(M)$.

Proof: It suffices to construct a splitting $s : \mathrm{Aut}(\varphi) \to N$. We may in an obvious way identify $\mathrm{Aut}(\Phi)$ with G the galois group of $K = \mathbb{Q}(\varsigma)$ over $\mathbb{Q}$. Clearly G acts on R. By letting G act trivially on the $p\mathbb{Z}$ summand of M, we can extend the action of G to all of M and thus get an imbedding of G in $\mathrm{Aut}(M)$, indeed in N. This gives the splitting s. □

Let U be the group of units in R and define a map

$$\theta : \{\pm 1\} \times U \to Z$$

by

$$[\theta(\epsilon, u)](n, r) = (\epsilon n, ur)$$

for $\epsilon \in \{\pm 1\}, u \in U, n \in p\mathbb{Z} \subset M$, and $r \in R \subset M$.

Lemma 7.2: *θ is an isomorphism.*

Proof: Clearly θ is an injection. Now if $\psi \in Z$, by iii) of Exercise 7.3,

$$\psi(n, r) = (\psi_1(n), \psi_2(r))$$

where both ψ_1 and ψ_2 are isomorphisms. Hence $\psi_1(n) = \pm n$. Thus we can let

$$\epsilon = \frac{\psi_2(n)}{n}$$

for any $n \in p\mathbb{Z}$ and ϵ will be well-defined and equal to ± 1.

Since ψ_2 commutes with the action of Φ on M and since $R = \mathbb{Z} \oplus \mathbb{Z}\varsigma \oplus \cdots \oplus \mathbb{Z}\varsigma^{p-2}$,

$$\begin{aligned}\psi_2(r) &= \psi_2(\sum a_i\varsigma^i) = \sum a_i\psi_2(\varsigma^i)\\ &= \sum a_1\psi_2(\varsigma^i 1) = \sum a_i\varsigma^i\psi_2(1)\\ &= r\cdot\psi_2(1)\end{aligned}$$

and the fact that ψ_2 is an isomorphism implies that $\psi_2(1)$ must be a unit. Hence $\psi = \theta(\epsilon, \psi_2(1))$. □

We will henceforth omit θ and simply identify $\{\pm 1\} \times U$ with Z. Thus we have determined N.

$$N = G\cdot(\{\pm 1\} \times U)$$

where G, the galois group of K, acts trivially on $\{\pm 1\}$ and as usual on U.

Now we turn to the computation of N_α. First we must describe which $\alpha \in H^2(\Phi; M)$ corresponds to π. It follows from Section 5 of Chapter IV that

$$H^2(\Phi; M) \approx H^2(\Phi; R) \oplus H^2(\Phi; \mathbb{Z}) = 0 \oplus \mathbb{Z}_p$$

and α can be taken to be any generator of this group.

We must now compute how $N = G\cdot(\{\pm 1\}\times U)$ acts on $H^2(\Phi; M)$. We can see how $G = \mathrm{Aut}(\Phi)$ acts by examining the following chain of *natural* isomorphisms:

$$H^2(\Phi; \mathbb{Z}) \approx \mathrm{Ext}_{\mathbb{Z}}(H_1(\Phi; \mathbb{Z}), \mathbb{Z}) \approx \mathrm{Ext}_{\mathbb{Z}}(\Phi; \mathbb{Z}) \approx \mathrm{Hom}_{\mathbb{Z}}(\Phi, \mathbb{Q}/\mathbb{Z}).$$

Thus $\mathrm{Aut}(\Phi)$ is mapped isomorphically onto $\mathrm{Aut}(H^2(\Phi; \mathbb{Z}))$, and this tells us how G acts on $H^2(\Phi; \mathbb{Z}) = H^2(\Phi; M)$.

Exercise 7.4: Show that $U \subset N$ acts trivially on $H^2(\Phi; M)$ (and thus is surely in N_α) and that $\{\pm 1\} \subset n$ acts by $-1\cdot\beta = -\beta$ for $\beta \in H^2(\Phi; M)$.

Lemma 7.3: *There is a split exact sequence*

$$1 \longrightarrow U \longrightarrow N_\alpha \longrightarrow \mathbb{Z}_2 \longrightarrow 1,$$

so $N_\alpha = \mathbb{Z}_2\cdot U$. Furthermore $g \neq 1 \in \mathbb{Z}_2$ acts on U by complex conjugation.

Proof: We have seen that $U \subset N_\alpha$. The only other way to get an element that leaves α fixed is to combine something in G with -1. Since G is of order $p-1$, it has a unique element of order 2, namely complex conjugation, and this induces $\beta \mapsto -\beta$ in $H^2(\Phi; M)$. Hence the element $\psi_{-1} \in N$ given by

$$\psi_{-1}(n,r) = (-n, \bar{r})$$

induces the identity on $H^2(\Phi; M)$, and ψ_{-1} is of order 2. Thus $N_\alpha = \{1, \psi_{-1}\} \cdot U$ as claimed. □

At this point it may seem as though we are almost through with the computation of $A(X)$ since $H^1(\Phi; M)$ is certainly easy to compute. However, the action of N_α/Φ on $H^1(\Phi; M)$ turns out to be somewhat difficult to describe. To describe this action we have to calculate all of the entries in the basic 9-diagram for π.

Our next task is the description of $\mathrm{Aut}^0(\pi) \subset \mathrm{Aut}(\pi)$. For $n \in \mathbb{Z}$, let $n' \in \{0, 1, \ldots, p-1\}$ be defined by $n \equiv n' (\mathrm{mod}\ p)$ and define

$$f(n') = \begin{cases} 1 + \varsigma + \cdots + \varsigma^{n'-1}, & \text{if } n' \neq 0; \\ 0, & \text{if } n = 0. \end{cases}$$

Lemma 7.4: *Define $\theta : R \to \mathrm{Aut}^0(\pi)$ by*

$$[\theta(r_0)](n,r) = (n, r + f(n')r_0) \tag{49}$$

for $n \in \varsigma$ and $r, r_0 \in R$. Then θ is an isomorphism.

Proof: First we must check that $\theta(r_0)$ is a homomorphism.

Exercise 7.5: Do it. (**Hint:** Recall that $\varsigma^p = 1$ and $1+\varsigma+\cdots+\varsigma^{p-1} = 0$.)

To see that $\theta(r_0) \in \mathrm{Aut}^0(\pi)$, we restrict it to $M = p\mathbb{Z} \cdot R$ and see that

$$[\theta(r_0)](pn, r) = (n, r + f(0)r_0) = (pn, r).$$

It is clear that θ itself is a monomorphism. It remains to see that θ is surjective. To do this we need i) of Lemma 1.2 which says that

$$\mathrm{Aut}^0(\pi) \approx Z^1(\Phi; M).$$

We must show that any $z \in Z^1(\Phi; M)$ is of the form $\theta(r_0)$. We have

$$\varsigma \cdot z(\varsigma^i) - z(\varsigma^{i+1}) + z(\varsigma) = 0,$$

so z is completely determined (inductively) by $z(\varsigma) \in M$.

Exercise 7.6: Show that $\Sigma \cdot z(\varsigma) = 0$ where $\Sigma = \sum_{\sigma \in \Phi}$.

Hence $z(\varsigma) \in R$. A comparison of the definition of $\theta(r_0)$ (Equation (49)) with Equation (5) shows that the automorphism induced by z is precisely $\theta(r_0)$. □

We now know that $\mathrm{Aut}^0(\pi) \approx R$ and $N_\alpha \approx \mathbb{Z}_2 \cdot U$ so the middle row of the basic 9-diagram is the exact sequence

$$0 \longrightarrow R \longrightarrow \mathrm{Aut}(\pi) \longrightarrow \mathbb{Z}_2 \cdot U \longrightarrow 1. \tag{50}$$

Lemma 7.5: *The sequence (50) splits so $\mathrm{Aut}(\pi) \approx (\mathbb{Z} \cdot U) \cdot R$. Furthermore U acts on R by multiplication and if g is the non-trivial element of $\mathbb{Z}_2$, then $g \cdot r = -\varsigma \bar{r}$ for $r \in R$.*

Proof: We construct a splitting $s : \mathbb{Z}_2 \cdot U \to \mathrm{Aut}(\pi)$ by

$$[s(g)](n, r) = (-n, \bar{r}) \text{ and } [s(u)](n, r) = (n, ur).$$

Exercise 7.7: i) Check that these equations define automorphisms of π.

ii) Show that we can extend the definition of s to the whole semi-direct product $\mathbb{Z}_2 \cdot U$. (**Hint:** Show that $s(\bar{u})s(g) = s(g)s(u)$.)

Now we have a splitting s, and we must verify the assertion concerning the action of $\mathbb{Z}_2 \cdot U$ on R.

Exercise 7.8: i) Show that $s(u)\theta(r)s(u)^{-1} = \theta(ur)$.

ii) Show that $s(g)\theta(r)s(g)^{-1} = \theta(-\varsigma\bar{r})$. (**Hint:** Show that $f(p - \imath) + \varsigma\bar{f}(\imath) = 0$.)

The assertion we want follows easily from this exercise. □

Theorem 7.2 : *The basic 9-diagram for the fundamental group of a p* *dimensional $\mathbb{Z}_p$-manifold X with $\mathfrak{A}(X) = 1$ is*

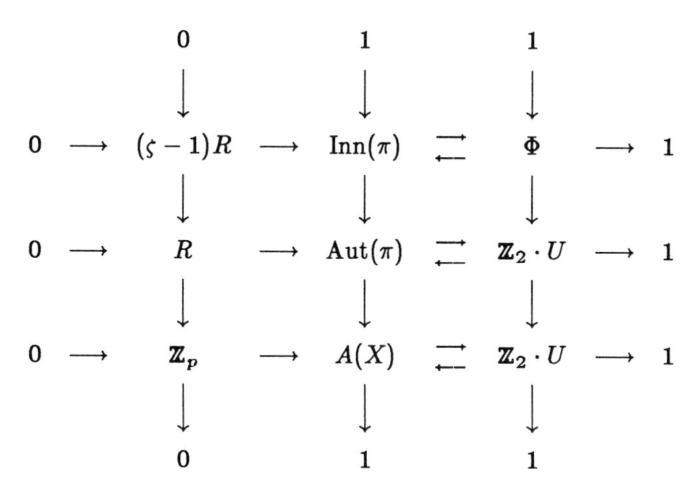

in which all rows split compatibly.

Proof: The only part that we have not already proved is the first column. Basically we are asserting that under the isomorophism $R \approx Z^1(\Phi; M)$ the subgroup $(\varsigma - 1)R$ goes into the subgroup $B^1(\Phi; M)$ which can be easily checked.

Exercise 7.9: Do it.

Exercise 4.5 of Chapter IV tells us that $R/(\varsigma - 1)R \approx Z_p$. □

Exercise 7.10: Show that the diagram indicates that the trace function $K \to \mathbb{Q}$ which sends x to $\sum_{\sigma \in G} \sigma \cdot x$ induces a specfic isomorphism $R/(\varsigma - 1)R \approx Z_p$.

Remark: The Dirichlet Unit Theorem ([77], page 145) says that U is the direct product of three groups: $\{\pm 1\}$, the group Φ generated by ς, and a free abelian group F of rank $(p-3)/2$. To see how $N_\alpha/\Phi = \mathbb{Z}_2 \cdot U/\Phi$ acts on $H^1(\Phi; M) \approx \mathbb{Z}_p$, observe that we know how $\mathbb{Z}_2 \cdot U$ acts on R, and this preserves $(\varsigma - 1)R$ so we can project modulo Φ. This action is impossible to describe directly because a basis for F is unknown.

It is known that the elements

$$\epsilon_i = \frac{\varsigma^i - 1}{\varsigma - 1} \qquad \text{for } i = 2, \ldots, \frac{p-1}{2}$$

generate a subgroup of maximal rank in F, and it is easily seen that ϵ_i acts by multiplication by i on $H^1(\Phi; M) \approx \mathbb{Z}_p$. However, if F' is the subgroup of F generated by the ϵ_i, the index of F' in F is the order of the ideal class group of the maximal real subfield $K[\varsigma + \bar{\varsigma}]$ which suggests that a basis for F may be hard to come by.

We now want to see how all of the above changes when X corresponds to an arbitrary $\mathfrak{A} = \mathfrak{A}(X) \in C_p$.

Exercise 7.11: Show that in this case $\pi = \pi_1(X)$ can be described as the semi-direct product $\mathbb{Z} \cdot \mathfrak{a}$ where $\mathfrak{a}$ is an ideal in $\mathfrak{A}$ and $n \cdot r = \varsigma^n r$ for $n \in \mathbb{Z}$ and $r \in \mathfrak{a}$.

First of all, Lemma 7.1 may not be correct in this situation. In general, G does not act on $\mathfrak{a}$, i.e. $x \in \mathfrak{a}$ does not imply $\sigma \cdot x \in \mathfrak{a}$ for $\sigma \in G$, so N/Z may not be all of $\mathrm{Aut}(\Phi) = G$. We can, however, still map U into Z and hence into N (actually into N_α) by virture of the following:

Lemma 7.6: *Let $f : \mathfrak{a} \to \mathfrak{a}$ be an isomorphism of R-modules. Then there is an element $u \in U$ s.t. $f(x) = ux\ \forall x \in \mathfrak{a}$.*

Exercise 7.12: Prove this lemma. (**Hint:** See Lemmas 22.2 and 22.4 on page 146 of [28] if you have difficulty.)

In the case of $\mathfrak{A}$ trivial (i.e. principal) we could combine the action of -1 with the element $g \in G$ that corresponds to complex conjugation to get an element in N that left α invariant. It would then appear that the only part of $\mathrm{Aut}(\Phi) = G$ we are interested in is g, the unique element of order two. The next lemma tells us that this is indeed the case and also tells us when there is an element in N_α that comes from this element of order two.

Lemma 7.7: *The image of N_α in $\mathrm{Aut}(\Phi)$ is either trivial or the subgroup of order two generated by g. The second alternative occurs iff $\mathfrak{A}(X) = \bar{\mathfrak{A}}(X)$.*

Proof: It follows from the basic 9-diagram that any element of N_α comes from an element $\psi \in \mathrm{Aut}(\pi)$. If we think of π as $\mathbb{Z} \cdot \mathfrak{a}$, the subgoup M is $p\mathbb{Z} \cdot \mathfrak{a}$. The subgroup $\mathfrak{a}$ of M is characterized by $\mathfrak{a} = \ker\Sigma$ where Σ is the automorphism of M induced by multiplication by $\sum_{\sigma\in\Phi} \sigma$. Since Σ is invariant under any automorphism of Φ, it follows that any automorphism of π maps $\mathfrak{a}$ onto itself. Thus ψ induces an automorphism of $\pi/\mathfrak{a} = \mathbb{Z}\cdot\mathfrak{a}/\mathfrak{a} \approx \mathbb{Z}$. Since $\mathbb{Z}$ has only one non-trivial automorphism, $1 \mapsto -1$, and since $(1,0) \in \pi$ projects onto a generator of $\Phi = \pi/M$, we can conclude that

ψ maps to the identity map in $\mathrm{Aut}(\Phi) = G$ or to the map that is the combination of multiplication by -1 and complex conjugation. This proves the first assertion.

Now the subgroup $\{(pn, 0) : n \in \mathbb{Z}\} \subset M$ can be characterized as the subgroup left fixed by the action of Φ, i.e. M^{Φ}. Hence any automorphism (e.g. ψ) of π maps this subgroup to itself. As we have just seen the only possibilities are

$$\psi(pn, 0) = (\pm pn, 0).$$

Claim: If the $+$ sign holds, then for some unique $u \in U$, we have

$$\psi(pn, \lambda) = (pn, u\lambda)$$

for $n \in \mathbb{Z}$ and $\lambda \in \mathfrak{a}$.

To see this, let $f : \mathfrak{a} \to \mathfrak{a}$ be the map induced by ψ, i.e. $\psi(0, \lambda) = (0, f(\lambda))\ \forall \lambda \in \mathfrak{a}$. Since

$$(1, 0)(0, \lambda)(-1, 0) = (0, 1 \cdot \lambda) = (0, \varsigma\lambda),$$

and since for some $s \in \mathfrak{a}$

$$\psi(1, 1) = (1, s),$$

we see that

$$f(\varsigma\lambda) = \varsigma f(\lambda) \qquad \forall \lambda \in \mathfrak{a}.$$

Now Lemma 7.6 says that there is an element $u \in U$ s.t. $f(\lambda) = u\lambda\ \forall \lambda \in \mathfrak{a}$ and the claim follows.

Claim: If the $-$ sign holds, then there is a unique $\gamma \in \mathbb{Q}(\varsigma)(= K)$ s.t. $\bar{\mathfrak{a}} = \gamma\mathfrak{a}$, and

$$\psi(pn, \lambda) = (-pn, \bar{\gamma}\bar{\lambda}) \qquad \forall \lambda \mathfrak{a}.$$

To see this let, $f : \mathfrak{a} \to \mathfrak{a}$ be as above, i.e. $\psi(0, \lambda) = (0, f(\lambda))\ \forall \lambda \in \mathfrak{a}$. Reasoning as above we conclude that this time

$$f(\varsigma\lambda) = (\bar{\varsigma})^{-1} f(\lambda) \qquad \forall \lambda \in \mathfrak{a}.$$

Let $\bar{f} : \mathfrak{a} \to \bar{\mathfrak{a}}$ be defined by $\bar{f}(\lambda) = \overline{f(\lambda)}$. Thus $\bar{f}(\varsigma\lambda) = \varsigma\bar{f}(\lambda)$, so we can apply the obvious extension of Lemma 7.6 to $\bar{f}$ to get $\gamma \in K$ s.t. $\bar{f}(\lambda) = \gamma\lambda\ \forall \lambda \in \mathfrak{a}$. Thus $\bar{\mathfrak{a}} = \gamma\mathfrak{a}$ and

$$f(\lambda) = \overline{\bar{f}(\lambda)} = \bar{\gamma}\bar{\lambda} \qquad \forall \lambda \in \mathfrak{a},$$

and the claim follows. Notice that the second claim shows that the second alternative implies that $\overline{\mathfrak{a}(X)} = \mathfrak{a}(X)$. It remains to show the converse. If $\overline{\mathfrak{a}(X)} = \mathfrak{a}(X)$, then $\exists \gamma \in K$ s.t. $\gamma\mathfrak{a} = \bar{\mathfrak{a}}$. Define $\psi : \pi \to \pi$ by

$$\psi(n, \lambda) = (-n, \bar{\gamma}\bar{\lambda}).$$

Exercise 7.13: Show that $\psi \in N_\alpha$ and maps to the map that is the combination of multiplication by -1 and complex conjugation.

This finishes the proof of the lemma. □

Notice that we did not use the first claim in the proof. We will use it now though.

Lemma 7.8: *In the current situation, the exact sequence*

$$0 \longrightarrow \mathrm{Aut}^0(\pi) \longrightarrow \mathrm{Aut}(\pi) \longrightarrow N_\alpha \longrightarrow 1$$

splits.

Proof: We describe a splitting $s : N_\alpha \to \mathrm{Aut}(\pi)$ by using the elements $u \in U$ and $\gamma \in K$ described in the two claims above. Recall that in the above proof we started with an element of N_α, so u and γ depend only on the image, say $\tilde{\psi}$, of ψ in N_α. If we are in the situation of the first claim, put

$$[s(\tilde{\psi})](n, \lambda) = (n, u\lambda).$$

If we are in the situation of the second claim, put

$$[s(\tilde{\psi})](n, \lambda) = (-n, \bar{\gamma}\bar{\lambda}).$$

Exercise 7.14: Show that $s(\tilde{\psi})$ is a well-defined automorphism and that s is a splitting. (**Hint:** Calculate.)

This finishes the proof of this lemma. □

Recall the following definitions from Section 8 of Chapter IV.

Definition 7.1: If $\mathfrak{A} \in C_p$ satisfies $\bar{\mathfrak{A}} = \mathfrak{A}$, we say that it is an *ambiguous ideal class.* An ideal $\mathfrak{a}$ s.t. $\bar{\mathfrak{a}} = \mathfrak{a}$ is called an *ambiguous ideal.* An ambiguous ideal class that contains an ambiguous ideal is said to be *strongly ambiguous.*

Remark: It follows from the remarks in Section 8 of Chapter IV that C_{29} contains an ambiguous class that is not strongly ambiguous. On the other hand, if the class number of $R_0 = \mathbb{Z}[\varsigma + \bar{\varsigma}]$ is not zero, then any ideal of R_0 can be lifted to R to yield a strongly ambiguous class.

Now we examine the exact sequence

$$0 \longrightarrow U \longrightarrow N_\alpha \longrightarrow N_\alpha / U \longrightarrow 1. \tag{51}$$

Lemma 7.9: *Suppose $\mathfrak{A}(X)$ is ambiguous. Then N_α/U is of order two and acts on U via complex conjugation. Furthermore this extension is split iff $\mathfrak{A}$ is strongly ambiguous.*

Proof: Recall that $u \in U$ is mapped into the map $\psi \in N_\alpha$ defined by $\psi(n, \lambda) = (n, u\lambda)$ for $n \in \mathbb{Z}$ and $\lambda \in \mathfrak{a}$. It follows from the second claim in the proof of Lemma 7.7, that there is a unique $\gamma \in K$ s.t. $\bar{\mathfrak{a}} = \gamma \mathfrak{a}$, and ψ_0 defined by

$$\psi_0(n, \lambda) = -n, \bar{\gamma}\bar{\lambda}$$

is the only element of n_α not in U.

Exercise 7.15: Show that the action of N_α/U on U is complex conjugation.

Since N_α/U is of order two, we have

$$H^2(N_\alpha/U; U) \approx A/B$$

where $A = \{u \in U : \bar{u} = u\}$ and $B = \{u \in U : \exists v \in U \text{ s.t. } u = v\bar{v}\}$.

Exercise 7.16: Show that the the coholomogy class of the extension (52) corresponds to the coset (in $H^2(N_\alpha/U; U)$) given by $(\psi_0)^2$.

Now

$$(\psi_0)^2(n, \lambda) = (n, \bar{\gamma}\gamma\lambda).$$

Hence $\bar{\gamma}\gamma \in U$ is the element whose coset describes the extension. Suppose $\mathfrak{A}$ is strongly ambiguous. Then we can take $\gamma = 1$ and clearly this is in the trivial coset, so the extension splits.

Conversely if $\bar{\gamma}\gamma = \bar{v}v$ for some $v \in U$, then we must have $\gamma \in R$, so $\bar{\mathfrak{a}} = \mathfrak{a}$, and $\mathfrak{A}$ is strongly ambiguous. □

We summarize our results in the following theorem:

Theorem 7.3 : *Let X be a $\mathbb{Z}_p$-manifold of dimension p. Then $A(X)$ is the semi-direct product $N_\alpha/\Phi\cdot\mathbb{Z}_p$ where N_α/Φ can be described as follows:*

i) If $\mathfrak{A}(X)$ is not ambiguous, then $N_\alpha/\Phi = U/\Phi \approx \{\pm 1\}\times$ a free abelian group of rank $(p-3)/2$.

ii) If $\mathfrak{A}(X)$ is ambiguous, then N_α/Φ contains U/Φ (as above) as a subgroup of index two. In the extension

$$0 \longrightarrow U/\Phi \longrightarrow N_\alpha/\Phi \longrightarrow N_\alpha/U \longrightarrow 1,$$

the quotient acts trivially on U/Φ. The extension is split iff $\mathfrak{A}(X)$ is strongly ambiguous.

Proof: Practically all of this has already been done. The only part left to examine is the passage from N_α to N_α/Φ in ii). Since Φ is of order p and N_α/U is of order 2, the map $H^2(N_\alpha/U;U) \to H^2(N_\alpha/U;U/\Phi)$ is an isomorphism. Furthermore the units U are known to be of the form $\varsigma^i r$ for some $r \in R_0 \subset \mathbb{R}$. Thus the action given by complex conjugation on U yields the trivial action of N_α/U on u/φ. □

Remarks: i) There is an amusing geometric consequence of an ideal class being strongly ambiguous. Starting with a $\mathbb{Z}_p$-manifold X with an ambiguous ideal class $\mathfrak{A}(X)$ one can construct a flat affine manifold that has a flat riemannian structure iff $\mathfrak{A}(X)$ is strongly ambiguous. See [21], page 491 for details.

ii) The next flat manifolds one might try to classify are the $\mathbb{Z}_2 \oplus \mathbb{Z}_2$-manifolds. Very little is known in this case. Complete results may be difficulty to obtain because $\mathbb{Z}_2 \oplus \mathbb{Z}_2$ has an infinite number of indecomposable integral representations.

iii) Actually there are some easier groups to try as holonomy groups, namely D_{2p} the dihedral groups of order $2p$ where p is an odd prime. The study of such D_{2p}-manifolds was initiated in [22] and in the last section of [21]. The approach in [22] is much like the one taken in Chapters III and IV. In the last section of [21], we note that a D_{2p}-manifold is covered by a $\mathbb{Z}_p$-manifold and the sole non-trivial deck transformation must be an affinity. The results of this section are then used to prove that the lowest dimension in which a D_{2p}-manifold can appear is $p+1$.

Our final remark is to point out yet again that the subject of Bieberbach groups and flat manifolds offers many opportunties to combine many different branches of mathematics in the solutions of many outstanding open problems.

Bibliography

Ambrose, W., and I. M. Singer

[1] A theorem on holonomy, *Trans. Amer. Math. Soc.* **75** (1953), 428.

Ankeny, N., S. Chowla, and H. Hasse

[2] On the class numbers of the maximal real subfield of a cyclotomic field, *J. reine angew. Math.* **217** (1965), 217.

Auslander, L.

[3] Examples of locally affine spaces, *Annals of Math.* **64** (1956), 255.

[4] On the Euler characteristic of compact locally affine spaces, *Comment. Math. Helvet.* **35** (1961), 25.

[5] The structure of compact locally affine manifolds, *Topology* **3** (1964), 131.

Auslander, L., and M. Kuranishi

[6] On the holonomy group of locally Euclidean spaces, *Annals of Math.* **65** (1957), 411.

Beauville, A.

[7] Variétés Kählериennes dont la première classe de Chern est nulle, *J. Diff. Geom* **18** (1983), 755.

Berger, M.

[8] Sur les groupes d'holonomie des variétés à connexion affine et des variétés riemanniennes, *Bull. Soc. Math. France* **83** (1953), 279.

Bieberbach, L.

[9] Über die Bewegungsgruppen der Euklidischen Raume I, *Math. Ann.* **70** (1911), 297.

[10] Über die Bewegungsgruppen der Euklidischen Raume II, *Math. Ann.* **72** (1912), 400.

Borevich, Z. I., and I. R. Shafarevich

[11] *Number Theory*, Academic Press, New York, 1966.

Brown, H., R. Bülow, J. Neubüser, H. Wondratschok, and H. Zassenhaus

[12] *Crystallographic groups of four-dimensional space*, Wiley, New York, 1978.

Buser, P.

[13] A geometric proof of Bieberbach's theorems on crystallographic groups, *L'Enseignment Math.* **31** (1985), 137.

Buser, P., and H. Karcher

[14] The Bieberbach case in Gromov's almost flat manifold theorem, *Global Differential Geometry and Global Analysis, Proc., Berlin 1979, Lectures Notes in Math.* **838**, Springer–Verlag, Berlin, 1981.

Cartan, E.

[15] Les groupes d'holonomie des espaces généralisés, *Acta Math.* **48** (1926), 1.

Cartan, H., and S. Eilenberg

[16] *Homological Algebra*, Princeton University Press, Princeton, 1956.

Charlap, L. S.

[17] Compact flat Riemannian manifolds I, *Annals of Math.* **81** (1965), 15.

Charlap, L. S., and C.–H. Sah

[18] Compact Flat Riemannian manifolds IV, mimeographed notes.

Charlap, L. S., and A. T. Vasquez

[19] Compact flat Riemannian manifolds II, *Amer. J. Math.* **87** (1965), 551.

[20] The automorphism group of compact flat Riemnnian manifolds, *Differential Geometrie im Grossen,* Math. Forschungsinstitut, Oberwolfache, 1971.

[21] Compact flat Riemannian manifolds III, *Amer. J. Math.* **95** (1973), 471.

[22] Multiplication of integral representations of some dihedral groups, *J. Pure and Applied Algebra* **14** (1979), 233.

Cheeger, J., and D. Gromoll

[23] The splitting theorem for manifolds of non-negative curvature, *J. Diff. Geom.* **6** (1972), 119.

Chevalley, C.

[24] *Theory of Lie groups I*, Princeton University Press, Princeton, 1946.

Conner, P., and F. Raymond

[25] Derived actions, *Proceedings of the second conference on compact transformation Groups II, Lecture Notes in Math.* **299**, Springer–Verlag, Berlin, 1972.

[26] Deforming homotopy equivalences to homeomorphisms in aspherical manifolds, *Bull. Amer. Math. Soc.* **83** (1977), 36.

Cornell, G., and L. Washington

[27] Class numbers of cyclotomic fields, to appear in *J. Number Theory*

Curtis, C. W., and I. Reiner

[28] *Representation theory of finite groups and associative algebras*, Interscience, New York, 1962

[29] *Methods of representation theory with applications to finite groups and orders I*, Wiley–Interscience, New York, 1981.

Edwards, H. M.

[30] *Fermat's last theorem*, Springer–Verlag, New York, 1977.

Faltings, G.

[31] Endlichkeutssätze für abelsche Varietäten über Zahlkörpen, *Invent. Math.* **73** (1983), 349.

Farrell, F. T., and W.-C.Hsiang

[32] Topological characterization of flat and almost flat riemannian manifolds $M^n (n \neq 3, 4)$, *Amer. J. of Math.* **105** (1983), 641.

Fried, D., and W. M. Goldman

[33] Three-dimensional affine crystallographic groups, *Advances in Math.* **47** (1983), 1.

Frobenius, G.

[34] Über die unzerlegbaren diskreten Bewegungsgruppen, *Sitzangsber. König. Preuss. Acad. Wiss. Berlin* **29** (1911), 654.

Giffen, C.

[35] Diffeotopically trivial periodic diffeomorphisms, *Invent. Math.* **11** (1970), 340.

Goldman, W. M., and M. W. Hirsch

[36] A generalization of Bieberbach's theorem, *Invent. Math.* **65** (1981), 1.

Gras, G.

[37] Critère de parité du nombre de classe des extensions abéliennes réeles de $\mathbb{Q}$ de degré impair, *Bull. Soc. Math. France* **103** (1975), 177.

Gromov, M.

[38] Almost flat manifolds, *J. Diff. Geom.* **13** (1978), 231.

Gruenberg, K. W.

[39] *Cohomological topics in group theory, Lectures Notes in Math.* **143**, Springer–Verlag, Berlin, 1970.

Hantzsche, W., and H. Wendt

[40] Driedimensionale euklidische Raumformen, *Math. Annalen* **110** (1935), 593.

Hasse, H.

[41] *Über die Klassenzahl abelscher Zahlkörper*, Akademie–Verlag, Berlin, 1952.

Heath-Brown, D. R.

[42] Fermat's last theorem for "almost all" exponents, *Bull. London Math. Soc.* **17** (1985), 15.

Hilton, P., G. Mislin, and J. Roitberg

[43] H-spaces of rank two and non-cancellation phenomena, *Invent. Math.* **16** (1972), 325.

Hopf, H.

[44] Zum Clifford–Kleinschen Raumproblem, *Math. Annalen* **95** (1926), 313.

Huppert, B.

[45] *Endliche Gruppen I*, Springer–Verlag, Berlin, 1971.

Husemoller, D., and J. W. Milnor

[46] *Symmetric bilinear forms*, Springer–Verlag, Berlin, 1973.

Janssen, T., A. Janner, and E. Ascher

[47] Crystallographic groups in time and space, *Physica (Netherlands)* **41** (1969), 541.

Jones, A.

[48] Groups with a finite nymber of indecomposable integral representations, *Michigan Math. J.* **10** (1963), 257.

Kobayashi, S., and K. Nomizu

[49] *Foundations of differential geometry I*, Wiley–Interscience, New York, 1963.

Kuiper, N. H.

[50] Sur les surfaces localement affines, *Colloque de Géométrie Différentielle, Strasbourg*, 1953, 79.

Kummer, E.

[51] Mémoire sur la théorie des nombres complexes composés de racines de l'unitié et de nombres entiers, *J. Math. Pures et Appl.* **16** (1851), 377.

Lang, S.

[52] *Linear Algebra*, Second Edition, Addison–Wesley, Reading, 1971.

[53] *Cyclotomic fields I* and *II*, Springer–Verlag, New York, 1979 and 1980.

[54] Units and class groups in number theory and algebraic geometry, *Bull. Amer. Math. Soc.* **6** (1982), 316.

Lehmer, D. H., and J. Masley

[55] Table of the cyclotomic class numbers $h^*(p)$ and their factors for $200 < p < 521$, *Math. Comp.* **32** (1978), 577 (microfiche suppl.).

Linden, F. van der

[56] Class number computations of real abelian number fields, preprint, Univ. of Amsterdam, 1980.

Mac Lane, S.

[57] *Homology*, Springer–Verlag, Berlin, 1963.

Massey, W. S.

[58] *Algebraic topology: an introduction*, Springer–Verlag, New York, 1967.

Milnor, J. W.

[59] Construction of the universal bundle I, *Annals of Math.* **63** (1956), 272.

[60] Groups which act on S^n without fixed points, *Amer. J. Math.* **79** (1957), 623.

[61] *Morse Theory*, Princeton University Press, Princeton, 1963.

[62] Whitehead torsion, *Bull. Amer. Math. Soc.* **72** (1966), 358.

[63] *Introduction to algebraic K–theory*, Princeton University Press, Princeton, 1971.

[64] On fundamental groups of complete affinely flat manifolds, *Advances in Math.* **25** (1977), 178.

Nazarova. L. A.

[65] Unimodular representations of the four group, *Dokl. Akad. Nauk. SSSR* **140** (1961), 1011 = *Soviet Math. Dokl.* **2** (1961), 1304.

Nelson, E.

[66] *Tensor analysis*, Princeton University Press, Princeton, 1967.

Nomizu, K.

[67] *Lie groups and differential geometry* , The Mathematical Society of Japan, Tokyo, 1956.

Northcott, D. G.

[68] *An introduction to homological algebra*, Cambridge University Press, Cambridge, 1960.

Reiner, I.

[69] Integral representations of cyclic groups of prime order, *Proc. Amer. Math. Soc.* **8** (1957), 142.

Ribenboim, P.

[70] *13 lectures on Fermat's last theorem*, Springer–Verlag, New York, 1979.

Rim, D. S.

[71] On projective class groups, *Trans. Amer. Math. Soc.* **98** (1961), 459.

Samuel, P.

[72] *Algebraic theory of numbers*, Hermann, Paris, 1970.

Schrutka von Rechtenstamm, G.

[73] Tabelle der (Relativ)-Klassenzahlen der Kreiskörper, deren ϕ-Funktion des Wurzelexponenten (Grad) nicht grösser als 256 ist, *Abh. Deutschen Akad. Wiss. Berlin, Kl. Math. Phys.* **2** (1964), 1.

Seah, E., L. Washington, and H. Williams

[74] The Calculation of a large cubic class number with an application to real cyclotomic fields, *Math. Comp.* **41** (1983), 303.

Shanks, D.

[75] The simplest cubic fields, *Math. Comp.* **28** (1974), 1137.

Stoker, J. J.

[76] *Differential geometry*, Wiley–Interscience, New York, 1969.

Washington, L.

[77] *Introduction to cyclotomic fields*, Springer–Verlag, New York, 1982.

Wolf, J.

[78] Discrete groups, symmetric spaces, and global holonomy, *Amer. J. Math* **84** (1962), 527.

[79] *Spaces of constant curvature*, second edition, Berkeley, 1972. (There are more recent editions of this book.)

Yau, S.–T.

[80] On Calabi's conjecture and some new results in algebraic geometry, *Proc. Nat. Acad. Sci. U.S.A.* **74** (1977), 411.

Index